Thomas Ring

Concrete Structures under Fire Loading

Thomas Ring

Concrete Structures under Fire Loading

From Experimental Characterization, Multiphase Modeling up to Structural Simulations of different Tunnel Geometries

Südwestdeutscher Verlag für Hochschulschriften

Impressum / Imprint
Bibliografische Information der Deutschen Nationalbibliothek: Die Deutsche Nationalbibliothek verzeichnet diese Publikation in der Deutschen Nationalbibliografie; detaillierte bibliografische Daten sind im Internet über http://dnb.d-nb.de abrufbar.
Alle in diesem Buch genannten Marken und Produktnamen unterliegen warenzeichen-, marken- oder patentrechtlichem Schutz bzw. sind Warenzeichen oder eingetragene Warenzeichen der jeweiligen Inhaber. Die Wiedergabe von Marken, Produktnamen, Gebrauchsnamen, Handelsnamen, Warenbezeichnungen u.s.w. in diesem Werk berechtigt auch ohne besondere Kennzeichnung nicht zu der Annahme, dass solche Namen im Sinne der Warenzeichen- und Markenschutzgesetzgebung als frei zu betrachten wären und daher von jedermann benutzt werden dürften.

Bibliographic information published by the Deutsche Nationalbibliothek: The Deutsche Nationalbibliothek lists this publication in the Deutsche Nationalbibliografie; detailed bibliographic data are available in the Internet at http://dnb.d-nb.de.
Any brand names and product names mentioned in this book are subject to trademark, brand or patent protection and are trademarks or registered trademarks of their respective holders. The use of brand names, product names, common names, trade names, product descriptions etc. even without a particular marking in this works is in no way to be construed to mean that such names may be regarded as unrestricted in respect of trademark and brand protection legislation and could thus be used by anyone.

Coverbild / Cover image: www.ingimage.com

Verlag / Publisher:
Südwestdeutscher Verlag für Hochschulschriften
ist ein Imprint der / is a trademark of
AV Akademikerverlag GmbH & Co. KG
Heinrich-Böcking-Str. 6-8, 66121 Saarbrücken, Deutschland / Germany
Email: info@svh-verlag.de

Herstellung: siehe letzte Seite /
Printed at: see last page
ISBN: 978-3-8381-3191-7

Zugl. / Approved by: Wien, TU, Diss., 2012

Copyright © 2012 AV Akademikerverlag GmbH & Co. KG
Alle Rechte vorbehalten. / All rights reserved. Saarbrücken 2012

Danksagung

Mein Dank gebührt meinen Arbeitskollegen am Institut für Mechanik der Werkstoffe und Strukturen (IMWS) an der Technischen Universität Wien (TU-Wien).

Besonders danken möchte ich meinen beiden Betreuern Univ.-Prof. Dipl.-Ing. Dr.techn. Roman Lackner und Dipl.-Ing. Dr.techn. Matthias Zeiml, die mich während meiner Dissertation stets mit Rat und Tat unterstützt haben und mit deren Ratschlägen alle Herausforderungen gelöst werden konnten.

Für die Unterstützung bei der Durchführung der Experimente im Rahmen meiner Arbeit möchte ich Wolfgang Dörner, Martin Reichel, Clemens Engler und Dr.phil. Roland Reihsner herzlich danken.

Ebenfalls bedanken möchte ich mich bei meinen unmittelbaren Arbeitskollegen im Labor des Instituts für Mechanik der Werkstoffe und Strukturen, Dipl.-Ing. Dr.techn. Aram Amouzandeh, M.Sc. Yiming Zhang, M.Sc. Shankar Shrestha, Dipl.-Ing. Artur Galek und Dipl.-Ing. Harald Moser, die mich auf meinem Weg der Dissertation begleitet haben.

Weiters sei die finanzielle Unterstützung des österreichischen Bundesministeriums für Verkehr, Innovation und Technologie (bm.vit) im Rahmen der KIRAS-Projekte (Österreichisches Förderungsprogramm für Sicherheitsforschung) 813794 und 824781 "Sicherheit von Hohlraumbauten unter Feuerlast" dankend erwähnt. Für die gute Zusammenarbeit möchte ich den Partnern der KIRAS-Projekte danken: Materialtechnologie Innsbruck (MTI) der Universität Innsbruck, Institut für konstruktiven Ingenieurbau der Universität für Bodenkultur Wien, Schimetta Consult ZT GmbH, Ingenieurbüro Dr. Lindlbauer, ZT Reismann, ÖBB-Infrastruktur Bau Aktiengesellschaft, ASFINAG

Autobahnen- und Schnellstraßen-Finanzierungs-Aktiengesellschaft, Wiener Linien GmbH & Co KG, ARGE Bautech, Forschungsinstitut der Vereinigung der österreichischen Zementindustrie.

Den wichtigsten Rückhalt während der Abfassung meiner Dissertation habe ich von meiner Familie und von meinen Freunden erhalten. Durch deren Unterstützung habe ich alle Herausforderungen im Rahmen meiner Dissertation gemeistert und auch privat den nötigen Ausgleich gefunden. Nochmals herzlichen Dank!

Kurzfassung

Im Rahmen der KIRAS-Projekte (Österreichisches Förderungsprogramm für Sicherheitsforschung) 813794 und 824781 "Sicherheit von Hohlraumbauten unter Feuerlast (2008-2012)" wurde der Brandfall in einem Tunnel untersucht. Der Beitrag der vorliegenden Arbeit zu diesem Projekt umfasste die experimentelle Charakterisierung des Verhaltens von Beton unter Brandeinwirkung (sowohl auf der Material- als auch auf der Strukturebene), die modellmäßige Beschreibung sowie die Berücksichtigung dieser Modelle im Zuge numerischer Simulationen. Die Arbeit gliedert sich in fünf Publikationen:

Publikation A ("Experimentelle Untersuchung der Verzerrungen von Beton und Zementstein unter hohen Temperaturen") beschreibt die experimentelle Untersuchung des Verformungsverhaltens an zylinderförmigen Probekörpern. Neben den radialen und axialen Verzerrungen bietet diese Publikation auch einen ersten Einblick in die Entwicklung der elastischen Materialparameter (E-Modul, Querdehnzahl) mit zunehmender Temperaturbelastung.

Publikation B ("Thermo-mechanisches Verhalten von Beton unter hohen Temperaturen: Von der mikromechanischen Modellierung zu Sicherheitsbetrachtungen von Tunnel im Brandfall") behandelt die Entwicklung der Verzerrungen von Beton unter kombinierter mechanischer und thermischer Beanspruchung. Mithilfe einer differentiellen Formulierung der Verzerrungen, die anhand der experimentellen Ergebnisse aus Publikation A validiert wurde, wird der Einfluss des modifizierten Materialverhaltens auf die Strukturantwort von Tunnelquerschnitten dargestellt.

In den **Publikationen C** ("Unterirdische Betonrahmen unter Brandbelastung: Teil I – Großbrandversuche") und **D** ("Unterirdische Betonrahmen unter

Brandbelastung: Teil II – Simulation der Großbrandversuche") werden die im Sommer 2010 durchgeführten Großbrandversuche in zwei Teilen beschrieben. Teil I beinhaltet die experimentellen Beobachtungen (Temperaturverteilungen, Verschiebungen, Verdrehungen, Rissentwicklung bis hin zum Abplatzverhalten). Teil II beschreibt die numerische Simulation der Großbrandversuche unter Berücksichtigung von verschiedenen Materialmodellen, der beobachteten Abplatzungen sowie des kombinierten mechanischen und thermischen Verformungsverhaltens von Beton aus Publikation B.

Abschließend wird in **Publikation E** ("Der Einfluss des Rechen- und Materialmodells auf die Strukturantwort bei der Simulation von Tunnel unter Brandbelastung") der Einfluss des gewählten Materialmodells auf die Berechnung von Tunnelstrukturen unter Feuerlast untersucht. Durch Vergleiche der Schnittgrößen und Verformungen werden Beurteilungen zwischen den jeweiligen Materialmodellen (z.B. linear-elastisches Materialverhalten + äquivalente Temperatur vs. elasto-plastisches Materialverhalten + nichtlineare Temperaturverteilung) möglich.

Die in dieser Arbeit präsentierten Ergebnisse sollen helfen, das Verhalten von Beton bei hohen Temperaturen besser zu verstehen und weiters eine Hilfestellung im Hinblick sowohl auf die Tunnelbemessung als auch auf die Sicherheitsbewertung von bestehenden Tunnel unter Brandeinwirkung bieten.

Abstract

In the framework of the KIRAS-projects (Austrian Security Research Program) 813794 and 824781 "Safety of underground structures under fire loading (2008-2012)", the effect of fire in tunnels was investigated. This work contributed to the research project by covering the experimental characterization and modeling of the behavior of concrete subjected to fire loading (on the material as well as on the structural scale) as well as consideration of these models within numerical simulations of tunnel cross-sections. The work consists of five publications:

Publication A ("Experimental investigation of strain behavior of heated cement paste and concrete") describes the experimental investigation of the deformation behavior of cylindrical specimens. Besides the radial and axial strains, first insight into the temperature-dependent evolution of the elastic material parameters (Young's modulus, Poisson's ratio) is given.

Publication B ("Thermo-mechanical behavior of concrete at high temperature: From micromechanical modeling towards tunnel safety assessment in case of fire") deals with the evolution of strains under combined mechanical and thermal loading. A differential strain formulation is introduced and validated by means of the experimental observations from Publication A. Finally, the influence of the proposed differential strain formulation on the structural response is illustrated by means of numerical simulations of tunnels.

In **Publications C** ("Underground concrete frame structures subjected to fire loading: Part I – large-scale fire tests") and **D** ("Underground concrete frame structures subjected to fire loading: Part II – re-analysis of large-scale fire tests"), the large-scale fire tests performed in summer 2010 are presented in two parts. Part I contains the experimental data (temperature distributions, deformations, rotations, crack development, and spalling behavior). Part II deals with the numerical simulation of these large-scale fire tests considering different material models, the experimentally-observed spalling depth and the combined mechanical and thermal strain behavior taken from Publication B.

Finally, in **Publication E** ("The influence of different computational and material models in simulation of the structural performance of tunnels unter fire loading") the influence of the material model on the simulation results of tunnels under fire loading is investigated. Comparison of stress resultants and deformations allows an assessment of the investigated material models (e.g., linear-elastic material behavior + equivalent temperature vs. elasto-plastic material behavior + nonlinear temperature distribution).

The results presented in this work should provide a better understanding of the processes taking place in concrete under high temperature, providing assistance to improve the design and safety assessment of tunnels subjected to fire loading.

Contents

Introduction   10

     Motivation 10

     Research objectives and methodology 12

     Outline of the thesis 17

     Contribution by the author 23

A   Publication 1   **27**

     A.1   Introduction 28

     A.2   Experiments 29

         A.2.1   Testing device 29

         A.2.2   Materials 32

     A.3   Results and discussion 33

     A.4   Concluding remarks 40

B   Publication 2   **43**

     B.1   Introduction 44

     B.2   Experimental observation 46

         B.2.1   Deformation under thermo-mechanical loading 47

         B.2.2   Behavior of siliceous material 49

     B.3   Micromechanical model 49

	B.3.1	Effective elastic properties	50
	B.3.2	Effective (free) thermal strain	53
B.4	Implementation		54
B.5	Finite-element implementation and numerical results		61
B.6	Concluding remarks		64
B.7	Appendix		66
	B.7.1	Effective prescribed strains in two-phase materials	66

C Publication 3 — 69

C.1	Introduction – Motivation	70
C.2	Experimental setup	72
	C.2.1 Geometric properties and loading	72
	C.2.2 Material	75
	C.2.3 Measurements	78
C.3	Experimental results and discussion	80
	C.3.1 Temperature measurements	80
	C.3.2 Deformation measurements	84
	C.3.3 Spalling and cracks	89
C.4	Concluding remarks	95

D Publication 4 — 97

D.1	Introduction	98
D.2	Numerical model	99
	D.2.1 FE-Model	99
	D.2.2 Boundary conditions	100
	D.2.3 Loading	100
	D.2.4 Material degradation	102

D.3 Simulation parameters 103

    D.3.1 Linear-elastic material behavior + equivalent temperature distribution 103

    D.3.2 Elasto-plastic material behavior + nonlinear temperature distribution 104

D.4 Numerical results 106

D.5 Concluding remarks 114

E Publication 5   117

E.1 Einleitung 120

E.2 Simulationsparameter 122

    E.2.1 Geometrie und Belastung des Rechtecksquerschnitts . . . 122

    E.2.2 Geometrie und Belastung des Gewölbequerschnitts . . . 124

    E.2.3 Thermische Belastung 124

    E.2.4 Materialmodelle 125

    E.2.5 Temperaturabhängigkeit der Materialparameter 127

E.3 Ergebnisse 128

    E.3.1 Rechtecksquerschnitt 129

    E.3.2 Gewölbequerschnitt 135

E.4 Fazit 139

E.5 Ausblick 141

Concluding remarks & Outlook   143

Bibliography   149

Introduction

Motivation

Fire accidents can seriously harm the structural safety of concrete structures. Especially in terms of critical underground facilities like tunnels, this can lead to increased danger and finally collapse of underground structures and – in the worst case – to loss of human life. During fire exposure, the concrete lining is damaged due to various processes (i.e., thermal degradation, spalling (see Figure 1), etc.). Generally, spalling happens during the first 30 min of thermal loading (high temperature rise up to 1200 °C within a few minutes), reducing the cross-sectional area of the tunnel lining and therefore considerably affecting the load-carrying capacity of the structure. In order to assess the safety of concrete structures subjected to fire loading, the aforementioned processes taking place in heated concrete must be taken into account.

In recent years, micromechanical concepts were employed to describe the behavior of heated concrete, considering chemical and physical processes at the respective scale of their occurrence. This step towards micromechanical modeling of building materials in general (see, e.g., [2, 42, 57, 58, 61, 62, 63, 85]), and of heated concrete, in particular (see, e.g., [14]), has drawn benefit from advances in both micromechanics and experimental characterization of finer-scale properties (see, e.g., [29, 30, 31, 59]), which has led to the development

Figure 1: (a) Failure of a specimen in small-scale experiments (test of cylindrical specimens subjected to combined mechanical and thermal loading), (b) final state of spalling after large-scale fire test exposing the reinforcement bars

of so-called multiscale models. Over the last years, great efforts were laid on the transfer of information from finer scales towards the so-called macroscale (upscaling), employing either analytical methods, based e.g. on continuum micromechanics (see, e.g., [2, 22, 55, 59, 60]), or numerical methods using the finite-element method or limit-state approaches (see, e.g., [9, 23]).

It is well known that the evolution of thermal strains of concrete subjected to combined mechanical and thermal loading depends on the history of loading (path dependence), which has been identified in numerous experiments [8, 16, 37, 72, 76]. In the literature, this effect is referred to as load induced thermal strains (LITS), occurring only during the first heating of concrete [37]. and depending on the actual stress level in concrete. In order to consider LITS within modeling of concrete under combined thermal and mechanical loading, various approaches can be found in the literature [50, 73, 74, 75]. As LITS reduces the thermal restraint of concrete, it may have a significant influence on the overall behavior of concrete under thermal loading.

As regards the simulation and design of concrete structures under high temperatures, the so-called equivalent temperature [39, 64, 78] is usually employed in engineering practice, simplifying the real non-linear temperature distribution in case of fire loading. This approach allows the use of beam elements for numerical simulations which are easy to handle and widespread in present design programs. This approach, however, has some drawbacks, such as the missing possibility to consider spalling and the assumption of linear-elastic material behavior. In order to perform more sophisticated numerical simulations, more realistic analysis tools are required, taking the real non-linear material behavior and effects like spalling into account.

Research objectives and methodology

The main goal of this thesis is the development of a numerical simulation tool being able to capture the structural behavior of reinforced concrete structures subjected to fire loading. This tool was developed within the framework of two KIRAS research projects (Austrian Security Research Program) 813794 and 824781 "Safety of underground structures under fire loading", aiming at an improved understanding of the behavior of concrete at high temperature in order to finally increase the safety of underground structures subjected to fire loading. Within this projects, a holistic approach was chosen, covering the three main parts of assessment of concrete structures subjected to fire loading:

- **Part 1 – Fire:**
  The first part deals with the computation of temperatures in tunnels in case of fire by means of CFD-simulations (see Figure 2) [4]. Hereby, different tunnel cross-sections (rectangular, arched) are investigated. The temperature distribution in the cross-section as well as in the longitudinal direction of the tunnel are determined, providing information for rescue teams as well

as design engineers (with the latter using the temperature loading of the structure within the structural analysis, see [4] for details).

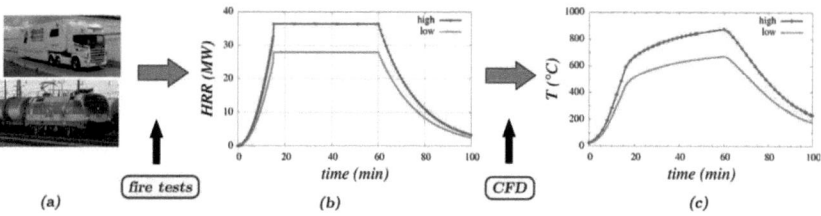

Figure 2: Application of CFD for the simulation of fire in enclosures: (a) type of fire load, (b) evolution of heat release rate (HRR), and (c) evolution of surface temperature [4]

- **Part 2 – Material:**

  The second part deals with the characterization of the material behavior of concrete under temperature loading. Hereby, the intrinsic permeability of various cement-paste and concrete mixtures is investigated in both the residual (up to 600 °C) and the hot state (up to 350 °C), where a considerable increase of the permeability of concrete with PP-fibers between 140 and 200 °C (comprising the melting temperature of PP-fibers) is observed (see [47] for details). In addition to the experimental investigation of the permeability of concrete, the mechanisms of spalling are numerically investigated in [84], modeling spalling as a combined thermo-mechanical and thermo-hydral process, depending on the stress level and the vapor pressure (see Figure 3). As regards the strain behavior of concrete under combined mechanical and thermal loading, the so-called "Load induced thermal strains" (LITS) are investigated [67].

- **Part 3 – Structure:**

  In the final part of the research projects, the behavior of concrete structures under fire loading is investigated. In addition to the simulation of different tunnel geometries subjected to fire loading [65], highlighting the influence of the material model on the structural response, the re-analysis of large-scale fire tests is performed (see Figure 4) [68, 69].

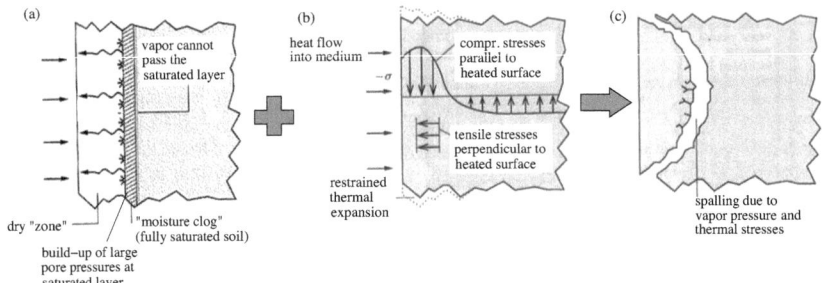

Figure 3: Illustration of spalling caused by (a) thermo-hydral processes and (b) thermo-mechanical processes [7, 12, 79]

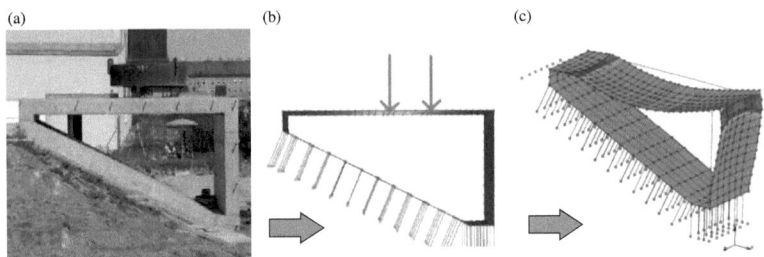

Figure 4: Large-scale fire tests: (a) experimental setup and (b,c) numerical model/results [68, 69]

Within this thesis, contributing mainly to Parts 2 and 3 of the aforementioned research projects, material models for concrete and modes of structural analyses are developed and validated by means of experiments:

Experiments

On the material scale, small-scale experiments on cylindrical specimens (with a diameter of 100 mm and a height of 200 mm) are performed. For this purpose, a test setup is developed, allowing the conduction of experiments under combined thermal and mechanical loading [70]. This test setup can be used for determination of the strain behavior as well as the elastic material properties of heated concrete. The displacements of the investigated specimens are measured with displacement transducers in radial as well as in axial direction. Steel bars are used to transfer the deformations from the hot test chamber to the outside of the testing device, as the displacement transducers cannot withstand the testing temperature (up to 800 °C). These steel bars are calibrated using a steel cylinder with known deformation behavior under high temperature. Knowing the displacements of the specimens under combined thermal and mechanical loading, the load induced thermal strains (LITS) can be determined. Since the mechanical loading can be varied during the experiment, small load changes are applied to determine Young's modulus and Poisson's ratio. In addition to the small-scale experiments, the material behavior of concrete under thermal loading was tested by means of real-scale experiments [66, 68, 69], where concrete frames (3 x 6 x 2 m) with different concrete mixtures (with and without PP-fibers) are heated and monitored as regards temperatures, deformation, rotation, spalling, and crack evolution. Finally, the collected experimental data is used to build up a proper validation basis for modeling and simulation of concrete and concrete structures at high temperatures.

Modeling of concrete

With the experimental basis at hand, a micromechanics-based model is proposed using continuum micromechanics for homogenization (upscaling). In a first approach, concrete is subdivided into aggregates, cement paste, and pore space. The micromechanics-based model is used to determine the effective elastic properties and the free thermal strain of concrete as a function of temperature. Furthermore, a differential strain formulation is proposed, accounting for the load-dependency of the thermal strains (LITS) by extending a model formulation for LITS originally proposed by Thelandersson [75]. The so-obtained model response shows good agreement with the experimentally-observed load-dependency of the thermal strain.

Structural simulations

Numerical simulations of the aforementioned large-scale experiments [66, 68, 69] as well as real-scale structures are performed (i.e., rectangular and arched tunnel cross-sections). The numerical simulations are performed using thick layered finite elements [71] within the numerical analysis tool MSC.Marc [48]. These layered finite elements (consisting of 105 layers) were employed, because:

- a significantly lower number of elements is required to model the tunnel cross-section as compared to, e.g., 3D-continuum elements, reducing the numerical cost,

- spalling of concrete layers can be considered by de-activation of the respective layers, and

- the non-linear temperature distribution and the material properties of concrete and steel can be assigned easily with sufficient accuracy.

Within the analyses, different material models are considered, ranging from simple linear-elastic material behavior using the equivalent-temperature concept up to more complex models using elasto-plastic material behavior and taking the (real) non-linear temperature distribution into account. The equivalent-temperature concept [39, 64, 78] was developed in order to linearize the non-linear temperature distribution over the thickness of a cross-section and to reduce the effort associated with numerical simulations of thermally-loaded structures. The principle of the equivalent-temperature concept can be explained as follows: The (real) non-linear temperature distribution is applied on a clamped beam, causing restraint-stresses which lead to internal stress resultants (normal force and bending moment). These stress resultants are used to determine the equivalent temperature, consisting of two parts, i.e., a constant temperature increase over the cross-sectional thickness T_m [°C] and a temperature gradient ΔT [°C/m]. This simplified temperature load can be considered within state-of-the-art analysis tools using beam elements (which are currently used in engineering practice). This approach has some drawbacks, e.g., disregard of force redistribution within a structure, leading to local over-/underestimation of internal stress resultants. In order to evaluate the accuracy of the equivalent-temperature concept during simulation of tunnel cross-sections, numerical simulations are performed and compared to results obtained from more complex models considering elasto-plastic material behavior in combination with the non-linear temperature distribution. Furthermore, the influence of the proposed micro-mechanical model as well as the influence of spalling is highlighted. Finally, validation of the numerical tool is performed comparing the structural response with experimental data collected throughout the mentioned large-scale experiments on reinforced concrete frames [68, 69].

Outline of the thesis

This thesis mainly deals with the simulation of concrete structures subjected to fire loading. Prior to the numerical simulations, however, experiments (on the material as well as the structural level) were carried out, providing (i) a better understanding of the behavior of concrete at high temperatures and (ii) a proper basis of the subsequent modeling attempts.

Starting with small-scale experiments in **Publication A**, the strain behavior of cylindrical specimens is experimentally characterized, highlighting the path-dependence of concrete when heated under mechanical loading. This path dependence, also referred to as "Load induced thermal strains" (LITS), was presented, e.g., in [75]. In Publication A, the specimens are tested at temperatures up to 800 °C with a heating rate of 1 °C/min. The dimensions of the specimens (diameter = 100 mm, height = 200 mm) in combination with the slow heating rate of 1 °C/min ensure that the specimens are heated almost uniformly. Temperature plots of the steel platens and tested specimens (reference specimen made of steel and concrete specimen) are shown in Figure 5, showing a negligible temperature difference within the steel specimen (3 °C) and a rather small difference of 15 °C within the concrete specimen at the maximum temperature in the course of the experiments (800 °C). The steel specimen is tested in order to calibrate the test setup. This calibration becomes necessary, because the deformations are measure with the help of steel bars pointing out from the heating chamber. Different specimens were tested, ranging from cement-paste to concrete specimens with and without PP-fibers. The main focus of the experiments is on the determination of the influence of mechanical loading on the evolution of thermal strain, where a reduction of thermal expansion with increasing mechanical loading is observed. The use of PP-fibers in the concrete mixture leads to an increase of pore space, when the PP-fibers melt (170 to 200 °C), which slightly reduces the thermal expansion of concrete. Taking advantage of the test setup which allows testing of specimens under combined mechanical and thermal loading, small load changes are applied periodically, giving access to the elastic material properties (Young's modulus, Poisson's ratio) as a function of temperature and mechanical loading. Again

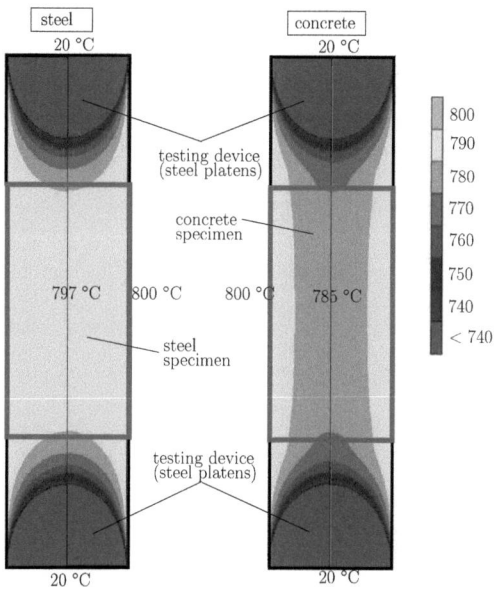

Figure 5: Numerically-obtained temperature distribution within the steel platens and the specimens (reference specimen made of steel and concrete specimen) at $T_{max} = 800$ °C

an influence of the applied mechanical load on the elastic material properties, especially Young's modulus, is observed. Increase of mechanical loading leads to a reduced development of micro-cracks and therefore to a more compact material at elevated temperatures, explaining the higher elastic properties.

Publication B deals with modeling of the experimentally-observed load dependency of thermal strains, as reported in Publication A. Within Publication B, a micromechanical model (consisting of aggregate, cement paste, pore space) is presented, giving access to the elastic material properties as well as the thermal strain of concrete. In a first approach, a total formulation of the thermal strain evolution based on a modified formulation proposed by Thelandersson [75] is applied, showing deficiencies when changes in the mechanical load-history of concrete are considered. As a remedy, a differential formulation of the strain evolution is proposed and validated by means of small-scale experiments. In the underlying small-scale experiments, the level of mechanical load level is changed during heating of the specimen. The publication is completed with numerical simulations of a rectangular tunnel cross-section illustrating the influence of the combined thermal and mechanical material behavior of concrete. The influence of LITS on the evolution of stresses in the cross-section is highlighted, showing a reduced stress built-up when LITS is considered. This reduced stresses lead to reduced bending moments in the whole structure. Furthermore, deformations are compared, showing smaller deformations in case LITS is considered.

Knowing the evolution of thermal strain with respect to mechanical loading (LITS) as well as the change of elastic material properties of concrete with respect to temperature, large-scale experiments are performed on concrete frames (6 x 3 x 2 m; test setup presented in Figure 6), see **Publication C**. These experiments are performed on four frames with different concrete mix-designs (two

Figure 6: Test setup in large-scale fire experiments (steel weight applied on closed fire chamber)

frames with and two frames without PP-fibers), carried out in order to provide the basis for validation of simulation tools. The shape of the large-scale experiment was chosen in order to determine the influence of fire on frame corners of rectangular tunnel cross-sections. The frames were loaded by steel weights, simulation an earth overburden on a tunnel cross-section. The fire load lasted for 180 min and reached temperatures in the inside of the combustion chamber of 1200 °C. A wide range of experimental data collected during the experiments like, e.g., temperature distributions, deformations, rotations of the frame corners, crack widths is presented. After the experiments the final spalling depth was recorded, showing higher spalling for the two concrete frames without PP-fibers. Additionally, tests on reference samples are performed, giving information about the compressive strength, Young's modulus, and permeability at certain temperatures. The complete set of data presented in this paper can be used for validation of models and simulations tools (spalling, heat transfer simulations, etc.).

A numerical tool based on [71] is used in **Publication D** in order to simulate the large-scale experiments presented in Publication C. The experimental data for deformations and rotations are used for validation of the model response. Different modes of material modeling are considered, starting with linear-elastic material behavior using the equivalent temperature concept up to elasto-plastic material behavior using the experimentally-determined non-linear temperature. Starting with linear-elastic material behavior, the complexity of the model increases continuously. Therefore, the influence of effects like spalling become evident. In the second step, elasto-plastic material behavior is introduced leading to more realistic results in comparison with the linear-elastic material behavior. The experimentally-observed extent of spalling is considered in a third step, highlighting the error in results in case spalling is neglected. Finally, consideration of LITS leads to the best agreement with experimental results, showing the need of considering LITS within the simulation of concrete structures exposed to fire loading.

Finally, the influence of the underlying material model on the structural response of tunnels is investigated in **Publication E**. Different cross-sections (rectangular, arched) are considered, starting with a simple material model (linear-elastic material behavior + equivalent temperature) up to more sophisticated material models (elasto-plastic material behavior + non-linear temperature distribution). Moreover, the influence of material properties taken from international standards (CEB or EC2-1-2) on the simulation results of concrete structures subjected to fire loading is investigated. As regards the underlying material behavior, the linear-elastic material behavior predicts the smallest deformations. Therefore, the state-of-the-art model approach cannot be chosen when realistic deformations of concrete structures subjected to fire loading should be obtained (see also Publication D). In addition to deformations, bending moments and normal forces are compared in Publication E highlighting the effect of force re-distribution when elasto-plastic material behavior is considered. As regards the underlying standards (CEB and EC2-1-2), the CEB approach produces higher thermal restraint leading to higher bending moments and larger deformations. This behavior is caused by the different evolution of Young's modulus with increasing temperature. Furthermore, the influence of the shape of the tunnel cross-section is determined, showing higher bending moments in frame corners of rectangular tunnel cross-sections.

Contribution by the author

Scientific contribution

- New experimental insights regarding the material behavior (strain behavior, Young's modulus, Poisson's ratio) of concrete subjected to thermal loading.
- A micromechanical model for heated concrete, covering the elastic properties and the free thermal strain.
- A differential formulation of LITS allowing consideration of the mechanical-load history of concrete subjected to thermal loading.
- An experimental basis including small-scale and large-scale experiments for validation of numerical tools to predict the behavior of concrete structures subjected to fire loading.

Engineering-oriented contribution

- The effect of different tunnel geometries (rectangular, arched) on the evolution of bending moments, normal forces, and deformations of tunnel structures with respect to fire loading.
- Assessment of different model assumptions (equivalent-temperature concept, non-linear material behavior, spalling, LITS), such as the equivalent-temperature concept leading to unrealistic small deformations of concrete structures.

This thesis is a cumulative work consisting of five publications, all written by the author of this thesis as first author. The following list gives more detailed information about the contribution of the author to each publication:

- **Publication A:** *Experimental investigation of strain behavior of heated cement paste and concrete* [70]

  This paper deals with experimental investigation of cement paste and concrete specimens subjected to mechanical and thermal loading. The design

and optimization of the test setup as well as the experiments and interpretation of the experimental data was carried out by the author himself.

- **Publication B:** *Thermo-mechanical behavior of concrete at high temperature: From micromechanical modeling towards tunnel safety assessment in case of fire* [67]

  Based on the findings of Publication A, existing micro-mechanics based models were adapted by the author and applied to modeling the behavior of heated concrete. The author compared the model response with the experimental results for validation purposes, finally studying the influence of LITS by means of structural simulations of a tunnel cross-section.

- **Publication C:** *Underground concrete frame structures subjected to fire loading: Part I – large-scale fire tests* [68]

  In this publication, the author summarizes the experimental data collected within large-scale fire tests. For details, the reader is referred to the respective technical report [66] (which was also written mainly by the author).

- **Publication D:** *Underground concrete frame structures subjected to fire loading: Part II – re-analysis of large-scale fire tests* [69]

  Using the experimental data from Publication C, the author investigated different model approaches (regarding the material behavior as well as the thermal loading) in order to determine the influence of spalling, LITS,... on the structural response, finally leading to the validation of the proposed non-linear structural model by comparison with experimental data (see Publication C).

- **Publication E:** *Der Einfluss des Rechen- und Materialmodells auf die Strukturantwort bei der Simulation von Tunnel unter Brandbelastung* [65]

Finally, the author performed structural simulations using the previously-developed numerical tool. Hereby, various numerical simulations of two different tunnel cross-sections were considered in order to determine the influence of thermal loading on deformations and the internal stress resultants (bending moment and normal force) of structures under fire loading. The numerical results led to identification of strengths and weaknesses of different numerical approaches as well as recommendations regarding the proper choice of the model. The respective conclusions provide the basis for preparation of an Austrian guideline covering this topic.

Publication A

Experimental investigation of strain behavior of heated cement paste and concrete [70]

Authored by Thomas Ring, Matthias Zeiml, Roman Lackner and Josef Eberhardsteiner

Published in Strain, 2012

The material degradation of concrete subjected to fire events has a severe influence on the load-carrying capacity of support structures. Spalling of concrete layers, exposing the reinforcement bars, and degradation of the material properties (Young's modulus, compressive strength, ...) may lead to significant damage of the reduced cross-section and, therefore, cause failure of the structure. In order to understand the stress build-up at the heated surface caused by thermal expansion due to fire loading, finally leading to damage and spalling of concrete, the strain behavior of cement paste and concrete exposed to combined thermo-mechanical loading is the focus of this work. Hereby, the evolution of thermal strains, Young's modulus, and Poisson's ratio with increasing temperature are investigated experimentally. For this purpose, the specimens are loaded uni-axially while the temperature is increased up to 800 °C. The obtained results provide the proper basis for the development of realistic material models, allowing more sophisticated simulations of structures exposed to fire.

A.1 Introduction

Fire loading of concrete structures (such as tunnels) can have a severe influence on the structural safety. In addition to the material degradation, spalling can be observed, resulting from the build-up of compressive stresses at the heated surface. The so-obtained combined thermo-mechanical loading was investigated, e.g., in [34, 73, 75], experiencing an increased amount of deformation under mechanical loading of heated concrete (see Figure A.1), requiring the intro-

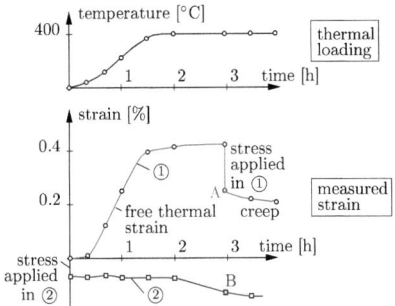

Figure A.1: Illustration of path dependence in case of combined mechanical and thermal loading [75]; the same thermal and mechanical loading applied in different order lead to different strains (compare points A and B)

duction of a new strain component referred to as load induced thermal strains (LITS). More recently [44], LITS was explained by strain incompatibilities of the constituents of concrete (cement paste and aggregates) under elevated temperatures, causing micro-cracking, which – under mechanical loading – are suppressed, giving rise to the notion of restrained thermal strain rather than LITS.

Accounting for the two main constituents of concrete controlling its performance, the behavior of both cement paste and concrete in investigated in this paper, considering combined thermo-mechanical loading. The experimental setup used in this work is presented in Section A.2, giving information about the thermo-mechanical testing device, the mix design for cement paste and concrete as well as the loading conditions. The experimental results (strains in axial and radial direction, Young's modulus, and Poisson's ratio) are presented in Section A.3, followed by concluding remarks in Section A.4, completing this work.

A.2 Experiments

A.2.1 Testing device

A radiant electric oven was used to apply thermal loading (see Figure A.2). The cylindrical oven was built around the mechanical testing device, allowing the conduction of tests under combined thermal and mechanical loading. Cylindrical specimens with a diameter of 100 mm and a height of 200 mm were

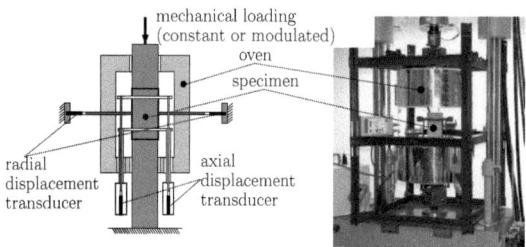

Figure A.2: Used device for thermo-mechanical testing of heated cement paste and concrete (see [24] for details)

tested. In order to monitor the deformations of the heated specimen in axial direction, steel rings were mounted to the specimen with steel bars transferring the deformation of the specimen towards the outside of the oven. In the radial direction, steel bars directly pointing from the specimen to the outside of the oven gave access to the radial deformation. Since the steel bars itself are subjected to temperature changes, the measured deformation is composed of the deformation of the specimen and the deformation of the steel bars. In order to determine the contribution of the latter, a reference test on a steel specimen with known thermal dilation was performed, allowing us to extract the deformation of the specimen from the monitored deformation history (see Figure A.3). In the course of the experiments, the specimens were subjected

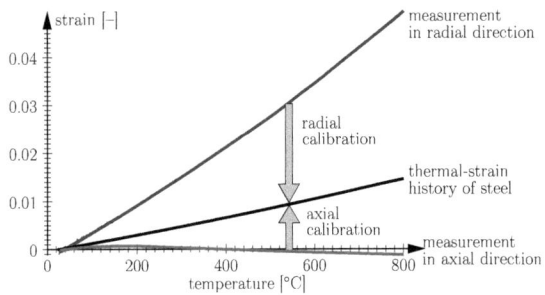

Figure A.3: Strain history in axial and radial direction of a steel specimen for calibration of the test setup

to uni-axial mechanical loading (with load level $s = 100 \cdot {}_a/f_{c,0}$ [%], where ${}_a$ is the applied axial stress and $f_{c,0}$ is the compressive strength at ambient temperature) which is applied prior to temperature loading and kept constant until the end of the experiment. Thereafter, thermal loading was applied up to 800 °C with a heating rate of 1 °C/min.

In order to identify the elastic properties of the heated specimens, the mechanical load was slightly decreased from the prescribed (constant) load level and subsequently reloaded at prescribed time intervals of 120 s (see Figure A.4). The continuous increase of temperature during change of the load level within the un-/reloading cycle of 15 s was found to be marginal, not affecting the determination of the elastic material properties.

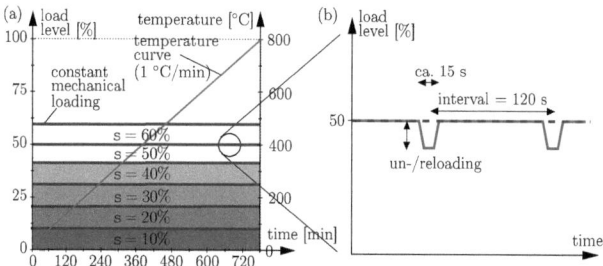

Figure A.4: (a) Thermo-mechanical loading of specimens and (b) history of load level considering un-/relaoding for determination of elastic material properties

Based on the monitored displacements in axial as well as radial direction during load change, Young's modulus and Poisson's ratio were determined from the so-obtained changes in axial and radial strain, reading

$$E(T,s) = \frac{\Delta_a}{\Delta_a} \qquad (A.1)$$

and

$$(T,s) = -\frac{\Delta_r}{\Delta_a}, \qquad (A.2)$$

where Δ_a refers to the change in axial stress associated with the slight decrease of mechanical loading. Due to the slight change in axial loading, the measured change in radial strain, Δ_r, is of small magnitude as compared to the change in axial direction giving more accurate results for Young's modulus than for Poisson's ratio.

A.2.2 Materials

Cement-paste consisting of CEM I 42.5 N with a water/cement ratio of 0.4 and a 28d-cylinder compressive strength of $f_{c,0}$ = 42.6 MPa was tested. In order to investigate an intermediate state between cement paste and concrete, additional experiments considering cement paste and filler (sand with grain size ≤ 0.25 mm) were performed (see Table A.1). The concrete mixture considered in the experimental program is given in Table A.2 (see also [24] for details). The

cement CEM I 42.5 N [kg/m^3]	781.6
additive (fly ash) [kg/m^3]	134.7
water [kg/m^3]	355.0
filler (siliceous) ≤ 0.25 mm [kg/m^3]	470.7
water/cement-ratio [–]	0.454
water/binder-ratio[†] [–]	0.399
fresh density [kg/m^3]	1742.0
28d-cylinder compressive strength [MPa]	47.7

[†] The additive is weighted by 0.8 [53].

Table A.1: Mix-design for specimens made of cement paste + filler

cement CEM I 42.5 N [kg/m^3]	290
additive (fly ash) [kg/m^3]	50
water* [kg/m^3]	185 / 190
polypropylene (PP) fibers* [kg/m^3]	0 / 1.5
siliceous aggregates [kg/m^3]	1859
fraction 0–4 mm [mass-%]	36
fraction 4–8 mm [mass-%]	17
fraction 8–16 mm [mass-%]	34
fraction 16–22 mm [mass-%]	13
aggregate mineralogy:	
quartz [mass-%]	68
feldspar [mass-%]	21
carbonate [mass-%]	11
water/cement-ratio* [–]	0.64 / 0.66
water/binder-ratio*,[†] [–]	0.56 / 0.58
initial density* [kg/m^3]	2384 / 2391
slump* [mm]	430 / 410
air content* [%]	1.0 / 2.5
28d-cylinder compressive strength* [MPa]	39.1 / 40.7

* Left value corresponds to concrete without PP-fibers, right value corresponds to concrete with PP-fibers.
[†] The additive is weighted by 0.8 [53].

Table A.2: Mix-design for concrete specimens without and with polypropylene-fibers (PP-fibers) [24]

considered aggregates consist of 89 % siliceous materials (68 % quartz and 21 %

feldspar). The underlying thermal-strain evolution of quartz is given in Figure A.5(a), indicating quartz transition at 573 °C. For T > 573 °C, the evolution of the thermal strain exhibits a plateau at 1.72 % (see Figure A.5(a)). The elastic properties of quartz as a function of temperature were determined in [43] using ultrasonic tests and are given in Figure A.5(b). Both free-thermal strain and elastic properties of quartz will be essential later on for the discussion of the obtained experimental results.

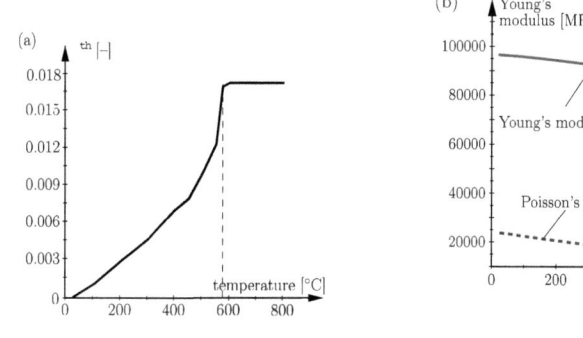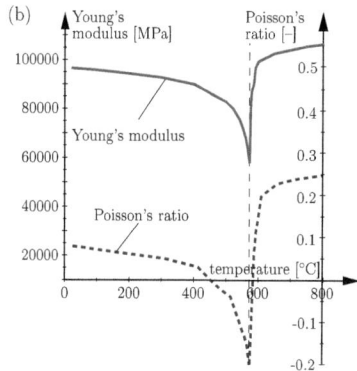

Figure A.5: Behavior of siliceous aggregates: (a) free thermal strain [33] and (b) Young's modulus and Poisson's ratio [43] as a function of temperature

A.3 Results and discussion

The evolution of strain in axial and radial direction for cement paste subjected to different levels of mechanical loading s ($s = a/f_{c,0} = 0, 5, 10, 20, 30 \%$, where $f_{c,0} = 42.6$ MPa, see Subsection A.2.2) is presented in Figures A.6(a) and (b), showing the load dependency of cement paste subjected to high temperature. In addition, the experimental results for unloaded specimens ($s = 0 \%$) from [73] are included, indicating lower compressive strains for the cement-paste mixture tested in [73]. With increasing load level, the compaction of cement

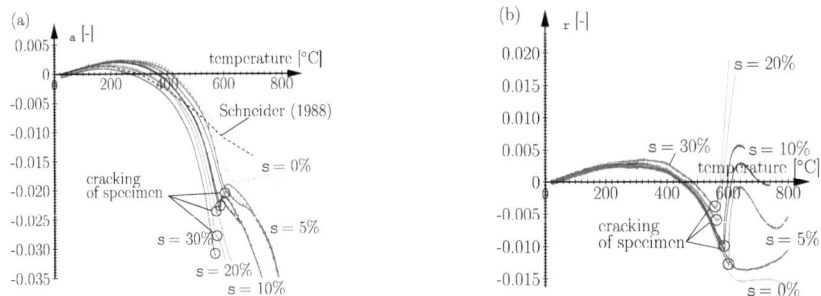

Figure A.6: Cement paste under combined thermal (up to 800 °C) and mechanical loading ($s = 100 \cdot {}_a/f_{c,0} = 0$ to 30 %, with initial compressive strength $f_{c,0} = 42.6$ MPa): evolution of (a) axial and (b) radial strain as a function of temperature

paste in axial direction increases, whereas the expansion in radial direction increases, especially at higher temperatures. While the behavior of strains below 500 °C is mainly driven by the degradation of CSH- and CH-phases (see [15, 77]), at temperatures between 500 to 600 °C a change of the evolution of strain is observed for $s > 5$ %, which is attributed to the development of macro-cracks (see Figure A.7) in axial direction of the specimen, caused by radial tensile stresses exceeding the (temperature-dependent) tensile strength. The development of these macro-cracks leads to a significant increase of deformation in the radial direction (see radial strain between 500 and 600 °C in Figure A.6(b)), starting at lower temperatures for higher level of mechanical loading. As regards determination of elastic properties and the strain behavior of heated cement paste, the strains measured after the formation of the macro-cracks cannot be considered. In case no mechanical load is applied during testing ($s = 0$ %), the evolution of the axial strain shows almost no further shrinkage for $T > 600$ °C, corresponding to decomposition of crystallized $CaCO_3$ replacing the earlier-observed degradation of CSH- and CH-phases (see, e.g., [15, 77]).

The evolution of Young's modulus of cement paste with temperature is presented in Figure A.8(a), indicating a rapid loss of stiffness in case of $s = 0$ %,

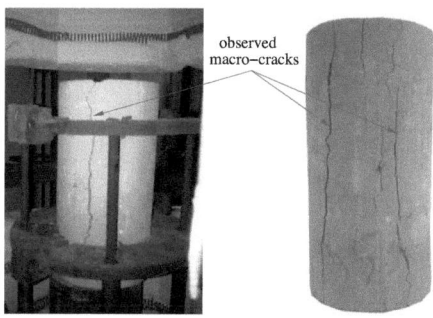

Figure A.7: Cement-paste specimens right after the experiment showing macro-cracks in axial direction

caused by both dehydration and the development of micro-cracks. Application

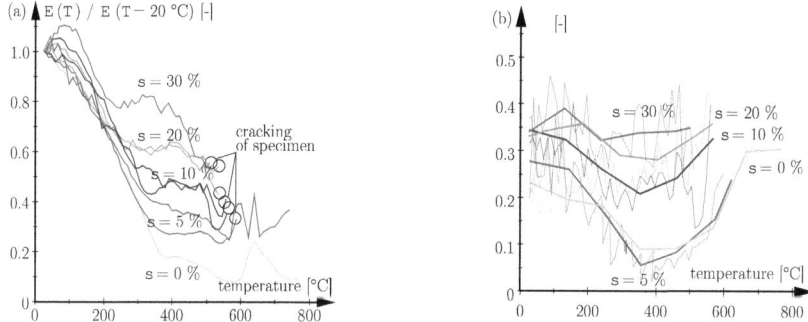

Figure A.8: Elastic parameters of specimens made of cement paste for different load levels ($s = 0, 5, 10, 20,$ and 30 %): evolution of (a) normalized Young's modulus and (b) Poisson's ratio as a function of temperature

of mechanical load leads to a reduction of microcracking and, thus, to higher values for Young's modulus at increased temperatures. In Figure A.8(b), the evolution of Poisson's ratio is presented. Since the applied load change for determination of Young's modulus and Poisson's ratio was rather small, the change in radial strain was in the range of the accuracy of the employed displacement-measuring device. Accordingly, the presented evolution of Poisson's ratio shows larger variations than the evolution of Young's modulus. Nevertheless, an increase of Poisson's ratio for cement paste under elevated temperatures for increasing mechanical loading is observed. For lower load levels ($s \leq 5\%$), Poisson's ratio was decreasing up to 400 °C, which is mainly explained by the occurrence of micro-cracks, weakening the cement-paste structure. Later on this trend is reversed and Poisson's ratio increases. For higher load levels ($s > 5\%$), the aforementioned behavior is less pronounced.

Figures A.9(a) and (b) show the evolution of axial and radial strain of the specimens made of cement paste + filler subjected to different levels of mechanical loading ($s = 100 \cdot {}_a/f_{c,0} = 0, 5, 10, 20, 30\%$, where $f_{c,0} = 47.7$ MPa, see Table A.1). Increasing the mechanical loading led to a rapid decrease of

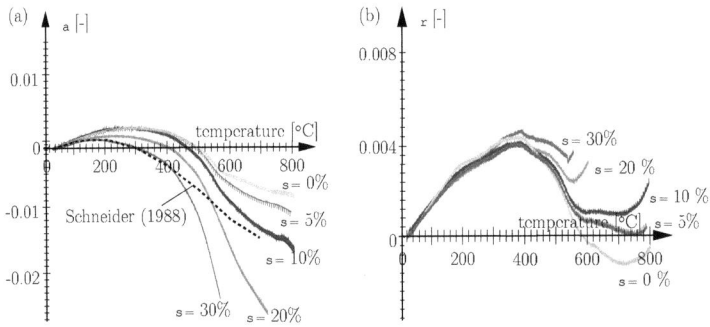

Figure A.9: Cement-paste + filler under combined thermal (up to 800 °C) and mechanical loading ($s = 100 \cdot {}_a/f_{c,0} = 0$ to 30 %, with initial compressive strength $f_{c,0} = 47.7$ MPa): evolution of (a) axial and (b) radial strain as a function of temperature

the axial deformation and later on to compressive strains (see Figure A.9(a)). As regards the radial strain (see Figure A.9(b)), less shrinkage deformation is obtained in consequence of mechanical loading, especially for $T > 400$ °C.

In Figure A.10(a), the evolution of Young's modulus of the specimens made of cement paste + filler is shown for different mechanical load levels ($s = 0, 5, 10, 20,$ and 30%). Young's modulus for the unloaded situation ($s = 0\%$)

is found to be significantly smaller than the modulus in case mechanical load is applied ($s \geq 5$ %). This is explained by the development of micro-cracks in addition to dehydration of the cement paste. Opening of these micro-cracks is suppressed for higher mechanical load levels, resulting in an increase of stiffness.

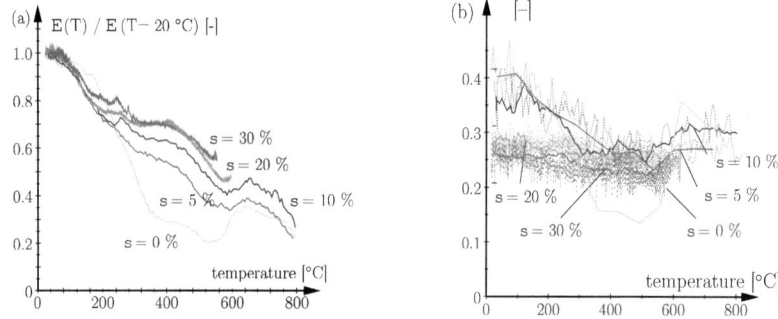

Figure A.10: Elastic parameters of specimens made of cement paste and filler for different load levels ($s = 0, 5, 10, 20$, and 30 %): evolution of (a) normalized Young's modulus and (b) Poisson's ratio as a function of temperature

The evolution of Poisson's ratio is presented in Figure A.10(b). The highest Poisson's ratio is obtained for the unloaded case, which can be explained by the small strain due to load change in combination with the presence of micro-cracks. For $s \leq 10$ %, Poisson's ratio increases at temperatures above 580 °C, related to the behavior of the siliceous filler material in combination with the degraded cement paste (micro-cracks). Application of mechanical load leads to a reduction of the formation of micro-cracks, and therefore to a more compact material, giving lower values for Poisson's ratios (e.g., $s = 20$ and 30 %). For $s = 20$ and 30 %, Poisson's ratio remains almost constant during heating, with a slightly lower Poisson's ratio for increased mechanical loading.

Finally, concrete specimens were subjected to combined mechanical and thermal loading. Figures A.11 and A.12 show the strain evolution for concrete without polypropylene-fibers (PP-fibers) and with PP-fibers in the axial (loaded)

and radial direction, respectively. The change in the strain evolution between 550 and 620 °C results from the quartz transition at 573 °C. The evolution of axial strain presented in Figures A.11(a) and A.12(a) decreases with increasing mechanical loading. At higher load levels ($s \geq 40$ %), the specimens fail before the final temperature of 800 °C is reached. As regards the corresponding radial strain (see Figures A.11(b) and A.12(b)), a continuous increase with

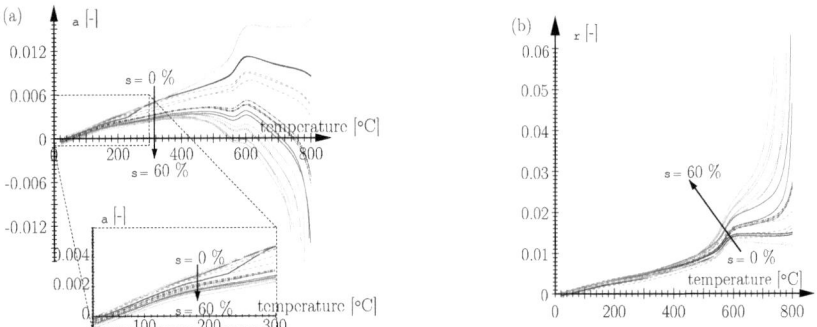

Figure A.11: Strain of concrete without PP-fibers under combined thermal (up to 800 °C) and mechanical loading ($s = 0$ to 60 % with $s = 100 \cdot {}_a/f_{c,0}$, where $f_{c,0} = 39.1$ MPa, see Table A.2): evolution of (a) axial and (b) radial strain as a function of temperature

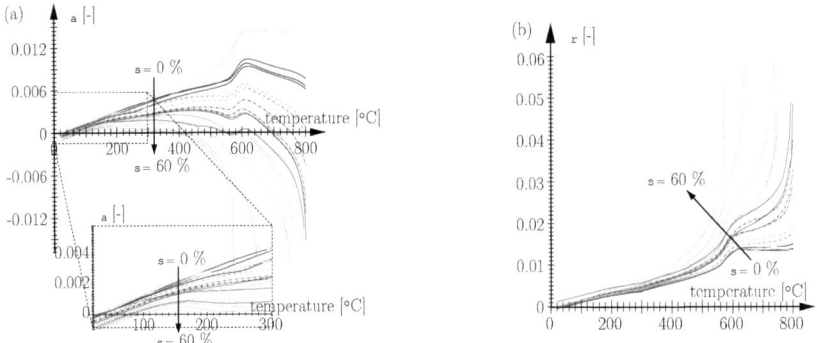

Figure A.12: Strain of concrete with PP-fibers under combined thermal (up to 800 °C) and mechanical loading ($s = 0$ to 60 % with $s = 100 \cdot {}_a/f_{c,0}$, where $f_{c,0} = 40.7$ MPa, see Table A.2): evolution of (a) axial and (b) radial strain as a function of temperature

increasing temperature is observed, which is more pronounced in case of higher mechanical loading. Comparison of Figures A.11(a) and A.12(a) shows the influence of PP-fibers (melting at approximately 170 °C), with the additional pore space increasing the effect of mechanical loading, leading to higher compaction with increasing load level (see Figure A.12(a)). Moreover, the pore space introduced by the PP-fibers leads to failure of the concrete specimens at lower temperatures as compared to the concrete without PP-fibers (see axial strains in Figures A.11(a) and A.12(a)).

In Figure A.13(a), the evolution of the normalized Young's modulus for concrete without PP-fibers subjected to different load levels (s = 0, 10, 20, 30, 40, and 50 %) is presented. For s = 0 %, the largest decrease in stiffness with increasing temperature is observed, with a minimum in the evolution of Young's modulus at the temperature of quartz transition (573 °C). Above that temperature up to 700 °C, the transition of quartz leads to a slight increase of the overall stiffness of concrete.

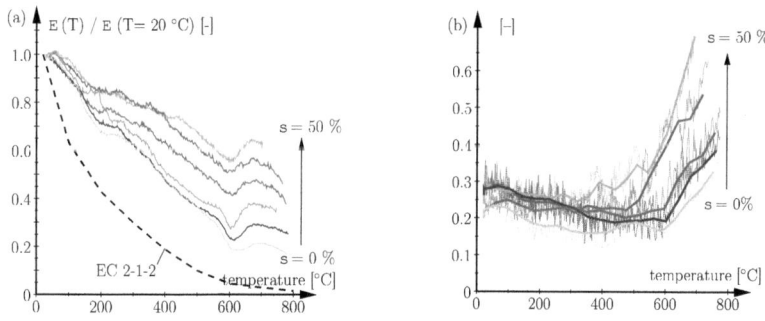

Figure A.13: Elastic parameters of specimens made of concrete without PP-fibers for di erent load levels (s = 0, 10, 20, 30, 40, and 50 %): evolution of (a) normalized Young's modulus compared to EC2-1-2 [20] and (b) Poisson's ratio as a function of temperature

The decrease of Young's modulus becomes smaller with increasing load level, which is explained by the reduced amount of micro-cracks, resulting from strain incompatibilities between aggregates and cement paste. In case of loaded concrete (with $s \geq 20$ %), the development of these micro-cracks is strongly reduced in comparison to non-loaded concrete ($s = 0\%$). The observed load-dependency of Young's modulus was also found in [34].

The influence of the load level ($s = 0, 20, 30, 40,$ and 50%) on the evolution of Poisson's ratio is presented in Figure A.13(b). Since the measurement accuracy in radial direction (small deformations) produces a certain variation in the results, the lines are not smooth. Up to $300\,°C$, Poisson's ratio is constant within a range of 0.2 to 0.3. For higher load level, a slight increase of Poisson's ratio was observed, which is more pronounced for higher temperatures. At temperatures above $573\,°C$ (quartz transition), Poisson's ratio is rapidly increasing, exceeding the elastic range for higher load levels (> 0.5), indicating inelastic effects like opening of additional microcracks within the cement-paste matrix.

A.4 Concluding remarks

The presented experimental results provided new insight into the behavior of concrete under combined thermo-mechanical loading as well as the behavior of its main constituents (cement paste, siliceous aggregates). The development of thermal strains was found to be highly dependent on the applied mechanical load, which is in good agreement to the open literature [34, 73]. The load dependency of the material properties was also encountered at the cement-paste scale, where the applied load seems to reduce the thermally-induced damage associated with microcracking. The presented experimental data are thought to provide a proper basis for understanding the behavior of concrete in the case of modeling attempts within the safety assessment of concrete structures subjected to fire. E.g., the additional load-induced thermal strain (LITS) accounting for the aforementioned load-dependency introduced by Thelandersson [75] may be reformulated from [67]

$$\varepsilon_a^{lits} = k\frac{\sigma}{f_{c,0}}\varepsilon_a^{th}(T) \quad \text{to} \quad \varepsilon_a^{lits} = \frac{\sigma}{f_c(T)}\varepsilon_a^{th}(T), \qquad (A.3)$$

where the compressive strength at ambient temperature, $f_{c,0}$, was replaced by the corresponding temperature-dependent compressive strength, $f_c(T)$ (see Figure A.14(a)). The parameter k, introduced in [75], depends on the type of loading (k = 2.35 for uniaxial loading, k = 1.7 for biaxial loading). Using the new formulation given in Equation (A.3), k becomes equal to 1.0 relating LITS directly to the actual level of loading, given by $\sigma/f_c(T)$. The response of the model proposed in Equation (A.3) (see [67] for details) shows good agreement to the experimental results presented within this paper (see Figure A.14(b)).

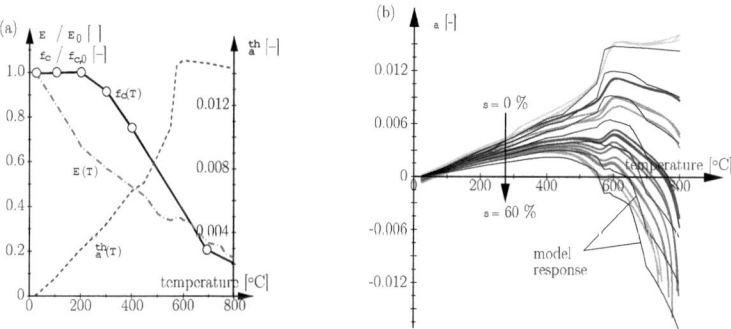

Figure A.14: Evolution of (a) input parameters (experimentally determined, see [67] for details) and (b) axial strain of concrete without PP-fibers (comparison of experimental results presented in this paper with $\varepsilon_a = \varepsilon_a^{el} + \varepsilon_a^{th} + \varepsilon_a^{lits}$, where $\varepsilon_a^{el} = \sigma/E(T)$ and ε_a^{lits} determined from Equation (A.3), with $f_c(T)$, $E(T)$, and $\varepsilon_a^{th}(T)$ taken from (a))

Publication B

Thermo-mechanical behavior of concrete at high temperature: From micromechanical modeling towards tunnel safety assessment in case of fire [67]

Authored by Thomas Ring, Matthias Zeiml and Roman Lackner

Published in Cement and Concrete Research, 2012

The realistic description of the behavior of concrete under high temperature and mechanical loading is of great importance, especially in terms of the safety assessment of concrete structures subjected to extreme events such as fires. In order to capture the complex chemical and physical processes in heated concrete, a micromechanical model taking the composite nature of concrete into account is presented in this paper. Based on experimental results obtained for cement paste and aggregate subjected to thermal/mechanical loading, a two-scale model formulated within the framework of continuum micromechanics is developed, giving access to the effective elastic and thermal-dilation properties of concrete as a function of temperature. These model-based properties are considered within a differential formulation of the underlying stress-strain law, accounting for the influence of mechanical loading on the thermal-strain evolution. The proposed micromechanical approach and its mode of implementation are validated by experimental results obtained from concrete specimens subjected to combined thermo-mechanical loading. Finally, the effect of the underlying model assumptions at the structural scale is illustrated by means of the safety assessment of underground support structures under fire attack.

B.1 Introduction

Concrete subjected to combined mechanical and thermal loading exhibits a certain path dependence explained by the dependence of physical processes on the actual stress state within the material (see [8, 16, 25, 50, 73, 74, 76]). This path dependence of heated concrete (highlighted in Figure B.1) is often related to the introduction of so-called load induced thermal strains (LITS). The main findings reported in the literature with respect to LITS are:

1. LITS is found only in concrete subjected to first thermal loading [36].

2. The rate of heating (ranging from 0.2 to 5 °C/min) and the water/cement ratio showed only minor influence on LITS [73].

3. The aggregate type has no significant influence in the development of LITS, linking LITS to processes taking place within the cement paste [36]

4. LITS is practically unaffected by the type of cement blend, suggesting that it takes place in a common gel or CSH structure [36].

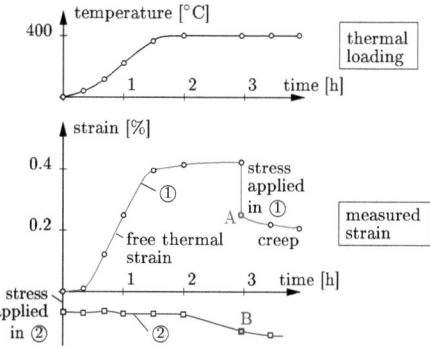

Figure B.1: Path dependence of combined mechanical and thermal loading according to [75]; the application of the same thermal and mechanical loading applied in different order lead to the same temperature and stress level ($T_{max} = 400$ °C and $\sigma = 0.45 \cdot f_{c,0}$) but to different experimentally-observed strains – compare points A and B

5. LITS seems to increase linearly with the applied stress level (see Figure B.2) [8].

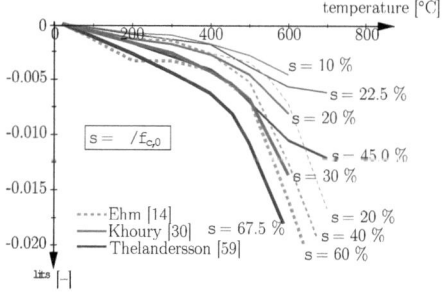

Figure B.2: Load dependency of LITS obtained from different experiments [16, 36, 75] ($s = \sigma/f_{c,0}$: level of loading)

Based on these experimental findings, several formulations for LITS can be found in the open literature, ranging from an approach to model LITS within creep of heated concrete [73] over considering LITS via empirical relations [74], to strain-rate formulations for LITS as proposed in [50, 75].

In recent years, micromechanics-based models for concrete have been published in the open literature (e.g., [14, 44]) taking the composite nature of concrete into account. On the one hand, these models were developed in order to identify the behavior of CSH at elevated temperatures [14] using nanoindentation. On the other hand, a multiscale model for determination of the effective stiffness of concrete at high temperatures was proposed in [44].

In the present work, recently published micromechanics-based models [42, 58] are adopted to the description of the change of elastic properties and the thermal dilation of heated concrete. For this purpose, experimental studies on concrete and cement-paste samples were conducted, the respective experimental results are presented in Section B.2. In Section B.3, the micromechanic-based model is presented with possible modes of implementation of the underlying stress-strain behavior which is discussed in Section B.4. The so-obtained formulations for consideration of the combined thermo-mechanical behavior of concrete and their effect within the analysis of concrete structures subjected to fire loading are highlighted in Section B.5. Concluding remarks are given in Section B.6.

B.2 Experimental observation

In addition to the existing experimental data available in the open literature, experiments were performed in order to assess the combined thermo-mechanical behavior as well as the elastic properties of concrete under temperature loading. The experiments were conducted in a radiant electric oven which is used to apply the thermal loading (see Figure B.3). The cylindrical oven is built around the mechanical testing device, allowing to perform tests under combined thermal and mechanical loading. The cylindrical specimens had a dimension of 100 mm in diameter and a height of 200 mm. In order to monitor the deformations

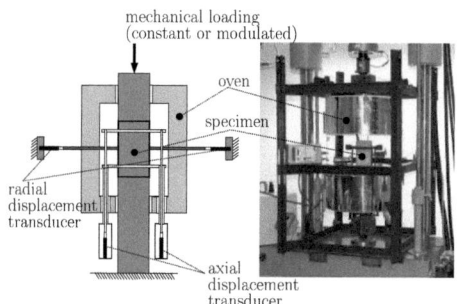

Figure B.3: Used device for thermo-mechanical testing (see [70] for details)

of the heated specimen, steel rings are mounted with steel bars transferring the deformation of the specimen to the outside of the oven (axial direction). In the radial direction, steel bars, directly pointing from the specimen to the outside of the oven, give access to the radial deformation. In the course of the experiments, the specimens are subjected to constant uniaxial loading and heated up to 800 °C with a heating rate of 1 °C/min.

In order to identify the elastic properties of the heated specimens, a modulated mechanical and steadily-increasing thermal load was considered within the test program (see [70] for details).

B.2.1 Deformation under thermo-mechanical loading

In Figure B.4, the evolution of strain in axial direction for cement paste subjected to different levels of mechanical loading ($s = 100 \cdot {}_a/f_{c,0}$ = 0, 5, 10, 20, and 30 %, where $f_{c,0}$ = 42.6 MPa) is presented, indicating the load dependency of deformations in case of increasing temperature loading. With increasing load level, the compaction of cement paste in axial direction and the expansion in radial direction increase, especially at higher temperatures. While the behavior of strains below 500 °C is mainly driven by the degradation of CSH – and CH -phases (see [15, 77]), at temperatures between 500 to 600 °C an abrupt

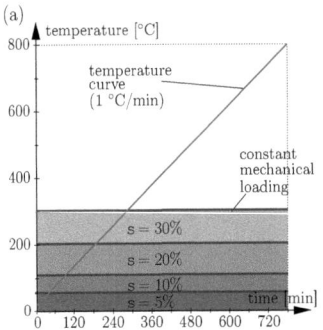 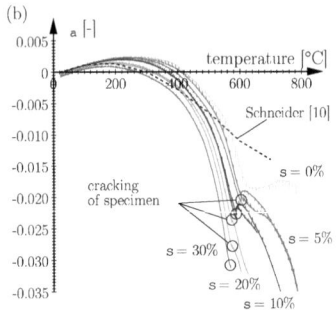

Figure B.4: Cement paste: (a) Temperature curve (1 °C/min) and mechanical loading ($s = 100 \cdot {}_a/f_{c,0} = 0$ to 30 %, with initial compressive strength $f_{c,0} = 42.6$ MPa); (b) evolution of axial strain as a function of temperature [70]

change of the evolution of strain is observed for $s > 5$ %, which is attributed to the development of macro-cracks in longitudinal direction of the cement-paste sample.

Figure B.5 shows the evolution of strain in axial and radial direction for concrete specimens under combined mechanical ($s = 100 \cdot {}_a/f_{c,0} = 0, 10, 20, 30, 40, 50,$ and 60 %, where $f_{c,0} = 39.1$ MPa) and thermal loading. The observed

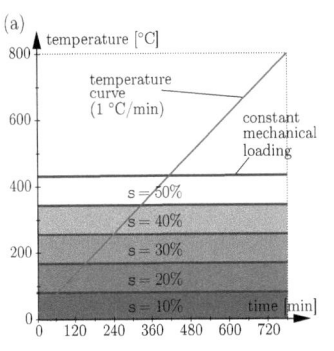 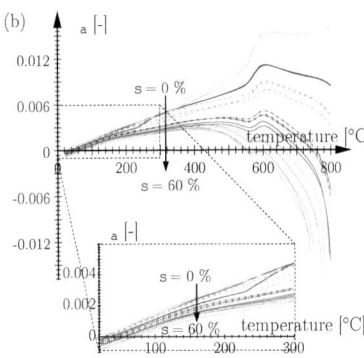

Figure B.5: Concrete: (a) Temperature curve (1 °C/min) and mechanical loading ($s = 100 \cdot {}_a/f_{c,0} = 0$ to 60 %, with the initial compressive strength $f_{c,0} = 39.1$ MPa); (b) evolution of axial strain as a function of temperature [70]

change in the strain evolution between 550 and 620 °C results from the quartz transition at 573 °C. The evolution of axial strain presented in Figure B.5(a) decreases with increasing mechanical loading. At higher load levels ($s \geq 40\,\%$), the concrete samples fail before the final temperature of 800 °C is reached. As regards the corresponding radial strain (see Figure B.5(b)), a continuous increase with increasing temperature is observed which is more pronounced in case of higher mechanical loading.

B.2.2 Behavior of siliceous material

Since concrete with a high content of siliceous aggregates (89 % in total, consisting of 68 % quartz and 21 % feldspar) was investigated in the previous subsection, the thermal strain behavior as well as the evolution of elastic properties is included in Section B.3, with the respective experimental results taken from [33, 43]. The thermal-strain evolution of quartz reported in [33] is shown in Figure B.9, indicating quartz transition at 573 °C. For $T > 573$ °C, the evolution of the thermal strain exhibits a plateau at 1.72 %. The elastic properties of quartz (Young's modulus and Poisson's ratio) as a function of temperature were determined in [43] using ultrasonic tests. Both free-thermal strain and elastic properties of quartz will be essential in the following section dealing with the micromechanical modeling of concrete behavior under combined thermo-mechanical loading.

B.3 Micromechanical model

In order to capture the influence of the constituents of concrete on the overall behavior, a micromechanical model is proposed, consisting of aggregates, cement paste, and pore space (see Figure B.6). Hereby, one portion of the air voids is already contained within the cement paste, while an additional portion of air voids is introduced by the mixing process of aggregates and cement paste (see Figure B.7(a)). Accordingly, the proposed micromechanical model comprises two scales in addition to the macroscale:

- At Scale I, cement-paste composite (pore space, hydration products) and additional pores introduced during the mixing process build up the material microstructure. At this scale, the experimentally-determined behavior for cement paste (see [70], for details) is considered.

- At Scale II, the aggregate phase is employed into the homogenized material of Scale I.

Within this micromechanical framework, both the effective elastic and thermal-dilation properties of heated concrete are determined using continuum micromechanics (based on Mori Tanaka [49], applied in [56]).

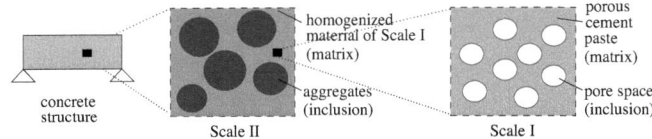

Figure B.6: Micromechanical model of concrete

B.3.1 Effective elastic properties

The effective shear and bulk modulus, G_{eff} and K_{eff}, are given as:

$$G_{eff} = \frac{\sum_r f_r G_r \left[1 + \left(\frac{G_r}{G_m} - 1\right)\right]^{-1}}{\sum_r f_r \left[1 + \left(\frac{G_r}{G_m} - 1\right)\right]^{-1}} \quad (B.1)$$

and

$$K_{eff} = \frac{\sum_r f_r K_r \left[1 + \left(\frac{K_r}{K_m} - 1\right)\right]^{-1}}{\sum_r f_r \left[1 + \left(\frac{K_r}{K_m} - 1\right)\right]^{-1}}, \quad (B.2)$$

with $r \in$ {porous cement paste (matrix), additional pore space (inclusion)} at Scale I and $r \in$ {homogenized material of Scale I (matrix), aggregates (inclusion)} at Scale II. The coefficients   and   are defined as

$$= \frac{3K_m}{3K_m + 4G_m} \quad \text{and} \quad = \frac{6(K_m + 2G_m)}{5(3K_m + 4G_m)}. \quad (B.3)$$

In Equations (B.1) to (B.3), the index m refers to the matrix phase, while   and   represent the volumetric and deviatoric part of the Eshelby tensor $, specialized for the case of spherical inclusions. Furthermore, f_r [-] refers to the volume fraction of the r-th material phase, which is determined from the concrete mix-design, with 1860 kg/m³ siliceous material, 330 kg cement / fly ash, and 185 kg water. Under the assumption of complete hydration, the initial volume fractions (prior to fire loading) for the investigated concrete mixture are set to $f_p/f_c/f_a$ = 0.05 / 0.25 / 0.7 (additional pore space (p) / porous cement paste (c) / aggregate (a), see Figure B.7).

As the material behavior (Young's modulus, Poisson's ratio) of porous cement paste is taken from the conducted experiments (see [70]), changes associated with dehydration are already considered in the cement-paste phase (see, e.g. [3, 38]).

The model response from Scale II, giving the effective Young's modulus (E_{eff}) for s = 0 %, is presented in Figure B.8, showing good agreement with experimental observations, especially for temperatures up to 200 °C. Table B.1 summarizes the evolution of the effective elastic properties obtained from the

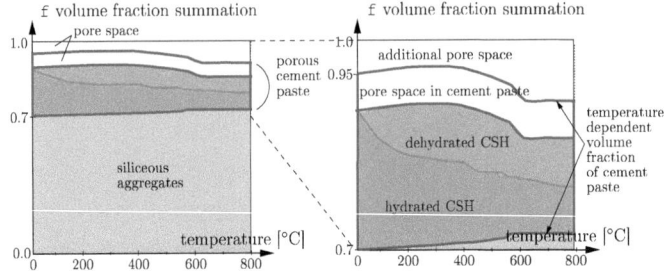

Figure B.7: Evolution of volume fractions for heated concrete

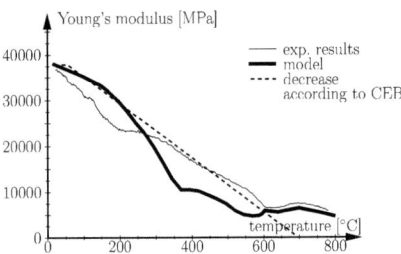

Figure B.8: Effective Young's modulus obtained from micromechanical model for unloaded concrete ($s = 0$ %) compared to experimental results (initial volume fractions: $f_p = 0.05$, $f_c = 0.25$, $f_a = 0.70$) and to decrease of stiffness according to CEB [11]

micromechanical model.

Temperature [°C]	K_{eff} [GPa]	G_{eff} [GPa]	ν_{eff} [–]	E_{eff} [GPa]
20	19.4	16.4	0.17	38.3
100	17.9	15.2	0.17	35.5
200	15.1	12.9	0.17	30.1
300	9.8	8.6	0.16	19.9
400	4.9	4.6	0.15	10.5
450	4.4	4.1	0.14	9.4
500	3.5	3.3	0.15	7.5
550	2.5	2.2	0.15	5.1
573	2.3	2.1	0.14	4.9
600	2.9	2.2	0.19	5.3
650	2.8	2.1	0.21	5.6
700	3.4	2.3	0.22	5.8
800	2.5	1.8	0.22	4.4

Table B.1: Effective elastic properties obtained from micromechanical model for unloaded concrete ($s = 0$ %)

B.3.2 Effective (free) thermal strain

When aggregates and cement paste are heated, they show a significant discrepancy in their thermal-dilation behavior (see Figure B.9). While siliceous aggregates are expanding during heating, cement paste is turning from expansion into shrinkage at 250 °C, which is explained by the continuous dehydration of cement paste [13, 73]. Using the morphology of the proposed micromechanical framework (Figure B.6), the effective thermal strain is given by (see Appendix A)

$$\varepsilon^{th}_{eff} = \varepsilon^{th}_{m} + (1 - f_m \langle A \rangle_{Vm}) \frac{K_i}{K_{eff}} (\varepsilon^{th}_{i} - \varepsilon^{th}_{m}), \quad (B.4)$$

where ε^{th}_{i} and K_i are the thermal strain and bulk modulus of the inclusion phase i (additional pore space at Scale I, aggregates at Scale II), respectively. Furthermore, the index m refers to the matrix phase at the respective scale. Figure B.9 contains the evolution of the effective thermal strain obtained from the proposed micromechanical model, showing excellent agreement with the experimental data. As indicated in Figure B.9, the free thermal strain is mainly driven by the behavior of the aggregates.

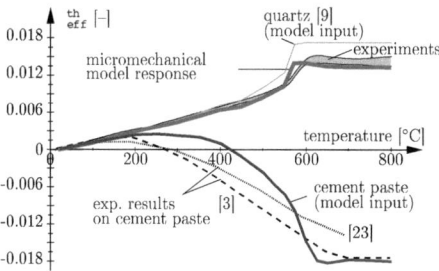

Figure B.9: Comparison of experimentally-obtained free thermal strain ($s = 0$ %) with prediction by micromechanical model together with experimentally-obtained thermal strain of the constituents (aggregates, cement paste), serving as model input

B.4 Implementation

In the open literature, the load-dependent part of thermal strains of concrete is often referred to as "Load Induced Thermal Strains" (LITS). LITS may be considered by an additional strain (see, e.g., [75]), reading

$$\varepsilon = \varepsilon^{el}(T, \sigma) + \varepsilon^{th}(T) + \varepsilon^{lits}(T, \sigma), \qquad (B.5)$$

where ε^{el} and ε^{th} represent the elastic strain and the free thermal strain (see ε^{th} in Figure B.10(a)), respectively. Commonly, the stress-dependence of LITS introduced in Equation (B.5) is considered by empirical relations, such as the Thelandersson-approach [75]:

$$\varepsilon^{lits} = k \frac{\sigma}{f_{c,0}} \varepsilon^{th}(T), \qquad (B.6)$$

where k is a parameter depending on the type of loading ($k = 2.35$ for uniaxial loading, $k = 1.7$ for biaxial loading [75]) and $\sigma/f_{c,0}$ accounts for the influence of the load level, giving a linear dependence of LITS on the applied stress (see Figure B.10(b)). For determination of LITS, the micromechanical model response for ε^{th} given in Figure B.10(a) is used.

Introducing the LITS-compliance tensor $\mathbb{D}^{lits}$, ε^{lits} given in Equation (B.6) may be formulated in a more general form, reading

$$\varepsilon^{lits} = \mathbb{D}^{lits}(T) : \boldsymbol{\sigma}. \qquad (B.7)$$

Combining Equations (B.5) and (B.7), the stress-strain law for heated concrete becomes

$$\boldsymbol{\sigma} = \mathbb{C} : \boldsymbol{\varepsilon}^{el} = \mathbb{C} : \left[\boldsymbol{\varepsilon} - \boldsymbol{\varepsilon}^{th} - \mathbb{D}^{lits} : \boldsymbol{\sigma}\right]. \qquad (B.8)$$

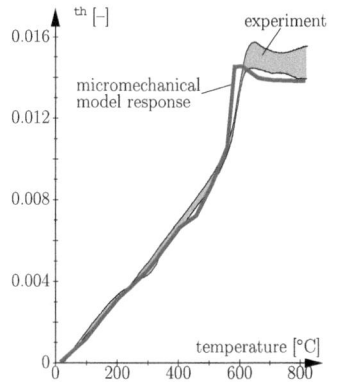
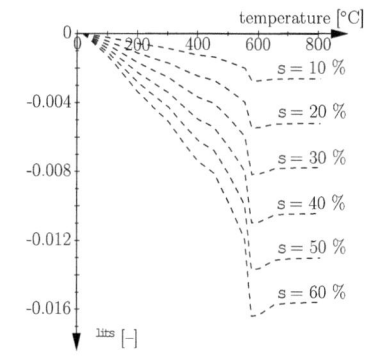

Figure B.10: (a) Evolution of free thermal strain of concrete, ε^{th}; (b) evolution of ε^{lits} for different load levels according to Equation (B.6) ($s = \sigma/f_{c,0} = 10, 20, 30, 40, 50,$ and 60%; $k = 2.35$; ε^{th} taken from Figure B.10(a))

Reformulation of Equation (B.8) gives

$$\varepsilon = \left(\mathbb{D} + \mathbb{D}^{lits}\right) : \sigma + \varepsilon^{th}, \qquad (B.9)$$

where $\mathbb{D} = \mathbb{C}^{-1}$ represents the elastic compliance tensor. Setting $\mathbb{D}^{lits} = k\mathbb{I}^{vol}\varepsilon^{th}/f_{c,0}$, the Thelandersson-approach given in Equation (B.6) is recovered. For the special case of axisymmetric conditions (axial and radial stress and strain components), the overall compliance tensor in Equation (B.9) becomes

$$\mathbb{D} + \mathbb{D}^{lits} = \frac{1}{E}\begin{bmatrix} 1 & -\nu \\ -\nu & 1 \end{bmatrix} + k\frac{\varepsilon^{th}}{f_{c,0}}\begin{bmatrix} 1 & 0 \\ 0 & 1 \end{bmatrix}, \qquad (B.10)$$

giving the axial strain ε_a in case of uniaxial loading, with the radial stress $\sigma_r = 0$, as

$$\varepsilon_a = \left(\frac{1}{E} + k\frac{\varepsilon^{th}}{f_{c,0}}\right)\sigma_a + \varepsilon^{th}. \qquad (B.11)$$

Comparison between the experimental results presented in Section B.2 with the results from Equation (B.11) reveals a significant deviation between the model response and experimental results, especially in the low-temperature regime (see Figure B.11).

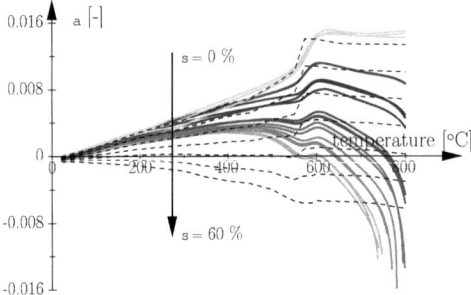

Figure B.11: Evolution of axial strain of concrete: Thelandersson-approach (Equation (B.6) with $k = 2.35$) compared with experimental results

In order to improve the agreement between model response and experimental data, the level of loading $\sigma/f_{c,0}$ in Equation (B.6) is reformulated, relating the stress to the *actual* compressive strength $f_c(T)$ (see Figure B.12). Accordingly,

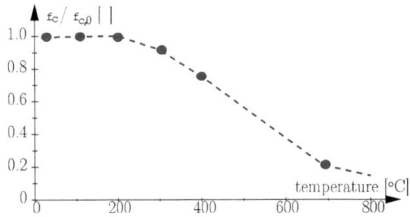

Figure B.12: Experimentally-obtained normalized compressive strength of concrete as a function of temperature [68]

$\mathbb{D}^{lits}$ in Equation (B.7) becomes

$$\mathbb{D}^{lits} = k \mathbb{I}^{vol} {}^{th}/f_c(T) , \qquad (B.12)$$

giving the axial strain in case of uniaxial loading as

$$a = \left(\frac{1}{E} + k\frac{^{th}}{f_c(T)}\right) a + {}^{th} . \qquad (B.13)$$

56

With this modification, the model response shows an improved agreement with the experimental results, especially in the low- and medium temperature regime (see Figure B.13). The temperature dependent compressive strength $f_c(T)$ was investigated in the literature [1, 36], highlighting a stress dependence of $f_c(T)$. For mechanically preloaded fire-exposed specimens the compressive strength was found to be higher than for mechanically-unloaded concrete. More recent investigations concerning the stress dependence of the compressive strength at 250 °C can be found in [54]. However, no stress dependence for the compressive strength $f_c(T)$ is considered in the proposed model up to now.

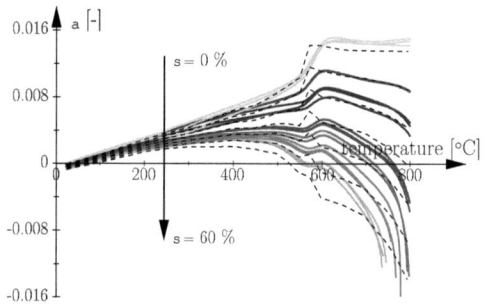

Figure B.13: Evolution of axial strain of concrete: modified Thelandersson-approach (Equation (B.13) with $k = 0.4$) compared with experimental results ($f_c(T)$ taken from Figure B.12)

So far, the model was applied and validated by means of experimental data only for constant mechanical loading. In real-life applications, however, the amount of mechanical loading certainly changes with time, e.g., in case of unloading, the elastic deformation (with $\varepsilon^{el} = \mathbb{D} : \boldsymbol{\sigma}$) vanishes. LITS deformations, on the other hand, account for the path-dependence of thermo-mechanical loading of heated concrete and must therefore remain. Accordingly, in contrast to the total formulation for the elastic strain $\varepsilon^{el} = \mathbb{D} : \boldsymbol{\sigma}$, a differential form is adopted for LITS, reading

$$d^{lits} = d\mathbb{D}^{lits} : \boldsymbol{\sigma}, \tag{B.14}$$

with the actual stress tensor $\boldsymbol{\sigma}$ affecting the differential change of LITS via $d\mathbb{D}^{lits}$. Replacing the differential changes in Equation (B.14) by finite changes within the time increment $n+1$, one gets:

$$\begin{aligned}\boldsymbol{\sigma}_{n+1} &= \mathbb{C}_{n+1} : \left[\varepsilon_{n+1} - \varepsilon_{n+1}^{th} - \left(\varepsilon_n^{lits} + \Delta\varepsilon^{lits}\right)\right] \\ &= \mathbb{C}_{n+1} : \left[\varepsilon_{n+1} - \varepsilon_{n+1}^{th} - \varepsilon_n^{lits} - \Delta\mathbb{D}^{lits} : \boldsymbol{\sigma}_{n+1}\right],\end{aligned} \tag{B.15}$$

where the incremental change of the LITS-compliance tensor is determined from $\Delta\mathbb{D}^{lits} = \mathbb{D}_{n+1}^{lits} - \mathbb{D}_n^{lits}$. Rewriting Equation (B.15) for the case of stress-driven situations (such as in case of axisymmetric uniaxial loading) gives

$$\boldsymbol{\varepsilon}_{n+1} = \mathbb{D}_{n+1} : \boldsymbol{\sigma}_{n+1} + \varepsilon_{n+1}^{th} + \varepsilon_n^{lits} + \Delta\varepsilon^{lits}. \tag{B.16}$$

with

$$\Delta\varepsilon^{lits} = \left(\mathbb{D}_{n+1}^{lits} - \mathbb{D}_n^{lits}\right) : \boldsymbol{\sigma}_{n+1}. \tag{B.17}$$

It can be seen in Figure B.14, that the agreement of the proposed incremental model with experimental data is equally good as the total formulation (see Figure B.13).

In order to validate the proposed differential formulation of LITS, experiments with changing load levels were performed and compared to the respective model response, considering both the modified total (Equation (B.13)) and the differential formulation (Equation (B.16)):

1. Within the first experiment, the level of loading is increased in four steps (see Figure B.15(a)). Starting from $s = 10\,\%$, the load level is increased in

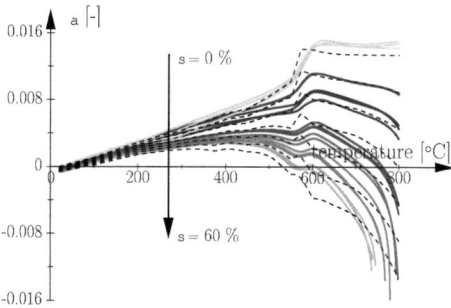

Figure B.14: Evolution of axial strain of concrete: differential formulation (Equation (B.16) with $k = 0.4$) compared with experimental results

three steps to 20, 30, and finally 40 %.

2. In the second experiment, the level of loading is first increased (from $s = 20$ to 40 %) and then reduced to zero loading ($s = 0$ %) (see Figure B.16(a)).

3. During the third experiment, the initial level of loading ($s = 40$ %) is decreased ($s = 10$ %) and finally increased ($s = 30$ %) (see Figure B.17(a)).

4. During the forth experiment, the initial level of loading ($s = 50$ %) is linearly decreased to $s = 0$ % between 400 and 670 °C and again linearly increased up to $s = 20$ % at 770 °C (see Figure B.17(a)).

For all experiments, the better agreement with the experimentally-obtained strain is found when using the proposed differential formulation (Equation B.16). The response of the modified total formulation (Equation B.13), strongly deviating from the experimental results, shows the largest error in case of mechanical unloading in the high-temperature regime.

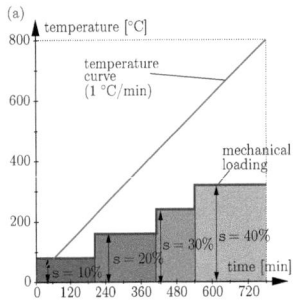 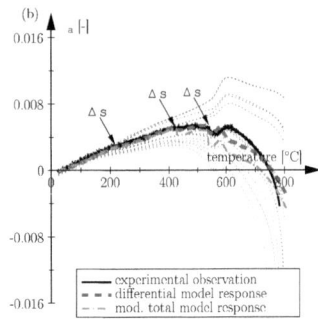

Figure B.15: Experiment 1: (a) Temperature curve (1 °C/min) and mechanical loading (from 10, 20, 30, to 40 %); (b) evolution of axial strain

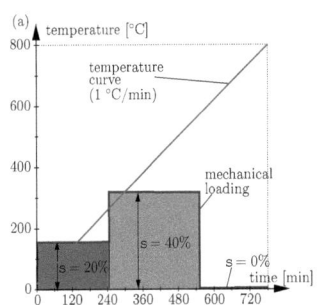 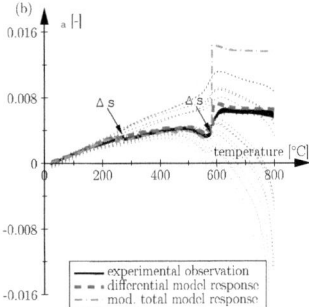

Figure B.16: Experiment 2: (a) Temperature curve (1 °C/min) and mechanical loading (from 20, 40, to 0 %); (b) evolution of axial strain

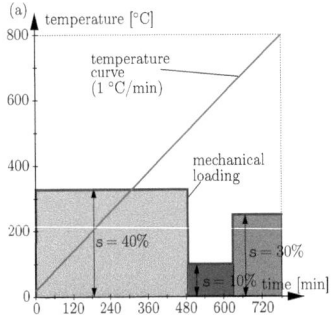 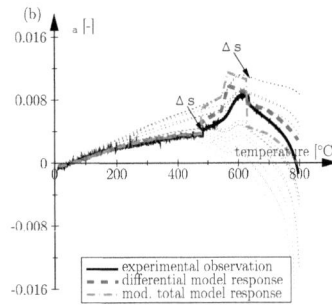

Figure B.17: Experiment 3: (a) Temperature curve (1 °C/min) and mechanical loading (from 40, 10, to 30 %); (b) evolution of axial strain

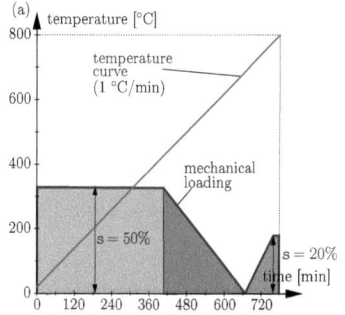
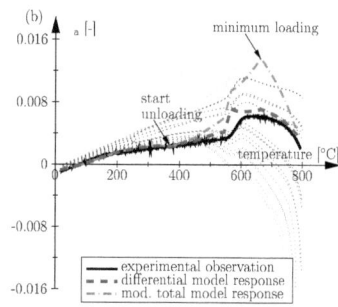

Figure B.18: Experiment 4: (a) Temperature curve (1 °C/min) and mechanical loading (linear decrease from 50 to 0 % and linear increase from 0 to 20 %); (b) evolution of axial strain

B.5 Finite-element implementation and numerical results

In this section, the differential formulation for the strain behavior of heated concrete outlined in the previous section is implemented into a finite element (FE) program [71]. In contrast to the stress-driven uniaxial stress situation encountered in the LITS experiments, strain increments are given by the underlying incremental-iterative solution procedure in nonlinear FE analysis, while the stress state σ_{n+1} at the end of the respective time increment needs to be determined. According to Equation (B.15), σ_{n+1} is given by

$$\sigma_{n+1} = \left[\mathbb{D}_{n+1} + \Delta\mathbb{D}^{lits}\right]^{-1} : \left[\varepsilon_{n+1} - \varepsilon_{n+1}^{th} - \overset{lits}{\varepsilon_n}\right]. \qquad (B.18)$$

In order to highlight the effect of the underlying LITS formulation, the proposed material model is used within the numerical analysis of a rectangular tunnel cross-section subjected to fire loading. The geometric properties of the considered tunnel cross-section are presented in Figure B.19(a). The thermal loading within the cross-section is obtained from a coupled thermo-hydro-chemical analysis [81], using the prescribed temperature loading within the tunnel shown in Figure B.19(b).

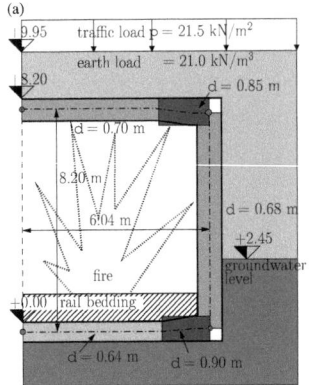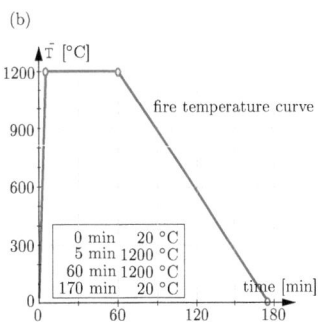

Figure B.19: Rectangular tunnel cross-section: (a) Geometric properties and (b) applied temperature loading within the tunnel (applied to side wall and ceiling of the frame)

In order to determine the influence of LITS on the structural response, three different material models are considered:

- Model 1 (no LITS): No consideration of LITS;
- Model 2 (TOT): Modified total formulation of LITS based on the Thelandersson-approach [75] (Equation (B.13));
- Model 3 (DIFF): Differential formulation of LITS (Equation (B.18)).

While LITS is considered only in case of compressive loading of concrete, the tensile stresses are limited in all considered models (Model 1 to 3) using a Rankine failure criterion, with $f_t(T) = 1/10 f_c(T)$ (according to [71]). The micromechanics-based Young's modulus (see Figure B.8) and free thermal strain (see Figure B.9) are employed. For the simulation of the reinforcement, a 1-D elasto-plastic material model was chosen, considering degradation of stiffness and yield-strength according to [20] (see [68] for details).

In the underlying fire scenario, a cooling phase is included (see Figure B.19(b)). During the heating phase, the evolution of the material properties is determined based on the temperature dependence shown in Figure B.12. During cooling the material properties (Young's modulus, compression/tensile strength) are dependent upon the maximum temperature reached. Since LITS was observed to take place during the first heating only [36], LITS is considered only during heating. In the course of cooling, no change of LITS takes place.

In Figure B.20, stress distributions at the top of the rectangular tunnel cross-section for different time instants (t_{fire} = 0, 20, 170 min) are presented. Model

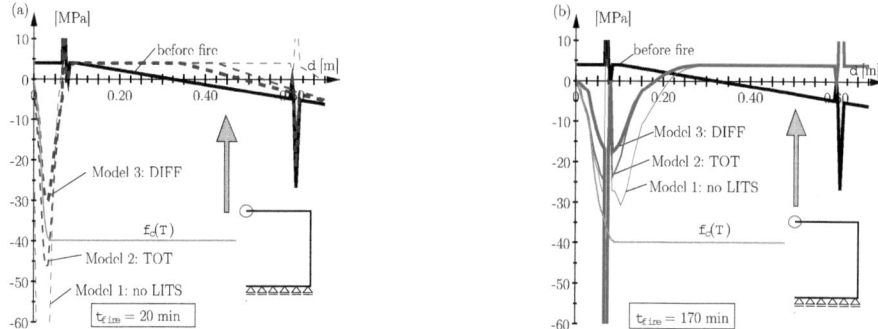

Figure B.20: Comparison of numerical results (Model 1 to Model 3): Stress distributions at (a) $t_{fire} = 20$ and (b) $t_{fire} = 170$ min

1 (no LITS) gives comparably high compressive stresses, even exceeding the compressive strength of concrete $f_c(T)$. Model 2 (TOT) results in a reduction of the stress built-up nevertheless, the stresses still exceed the compressive strength of concrete. Finally, Model 3 (DIFF) further reduces the compressive stresses which stay below the temperature-dependent compressive strength.

Figure B.21 shows the deformation of the tunnel cross-section, with the deformation history in the symmetry axis at the top of the tunnel given in Figure B.21(a) and deformation patterns of the whole cross-section given in Figure B.21(b). The largest restraint occurs for Model 1 (no LITS), resulting in large regions with plastic deformations within the reinforcement. On the other

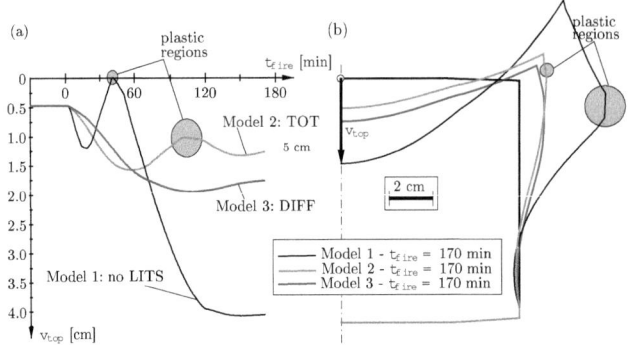

Figure B.21: Comparison of numerical results (Model 1 to Model 3): (a) deformation history at top of the tunnel; (b) deformation pattern of the whole cross-section at $t_{fire} = 170$ min

hand, the model response for Model 3 (DIFF) shows no plastic deformations of the reinforcement at all since the stress built-up due to thermal loading is considerably reduced by LITS, resulting in less loading of the concrete and, thus, the reinforcement bars.

Finally, the influence of the different models on the evolution of bending moments is presented in Figure B.22. In case of Model 3, the thermally-induced bending moments are reduced, resulting from lower stresses within the cross-section. On the other hand, totally neglecting the effect of LITS leads to the highest bending moments. For Model 3, the maximum bending moment at $t_{fire} = 170$ min is reduced (see Figure B.22), indicating – based on the more realistic material model for heated concrete – a higher safety of the underlying tunnel design.

B.6 Concluding remarks

Within this work, a differential strain formulation allowing the description of the path dependence of the strain behavior of concrete (LITS) in case of combined

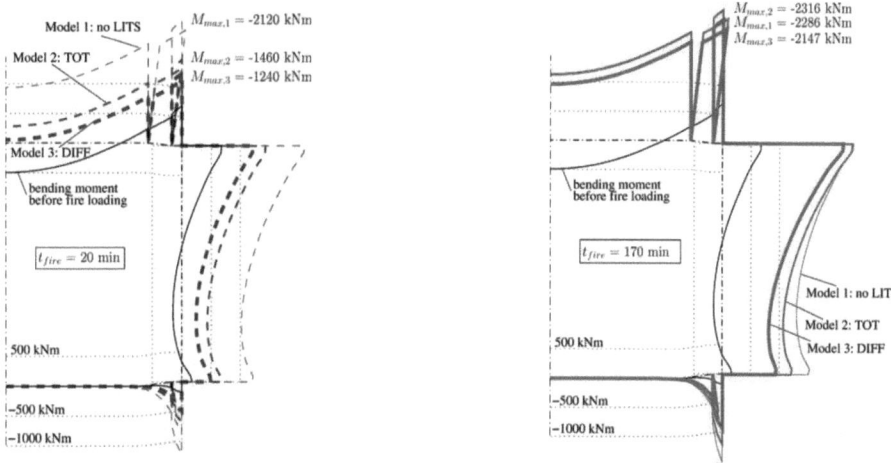

Figure B.22: Comparison of bending-moment distribution obtained with Model 1 to Model 3 at (a) $t_{fire} = 20$ min and (b) $t_{fire} = 170$ min

thermo-mechanical loading is proposed. The underlying material parameters for concrete (Young's modulus, Poisson's ratio, free thermal strain) are determined using a micromechanical model. Based on experimental observations highlighting the influence of level of loading on the strain behavior of heated concrete, the developed material model was validated. Finally, the effect of different modes of consideration of LITS on the structural response of a rectangular tunnel cross-section was assessed. Based on the obtained results, the following conclusions can be drawn:

- **Effect of plasticity:** The introduction of the differential formulation to consider LITS reduces compressive stresses induced by thermal restraint within concrete which then remain below the respective ultimate compressive strength. In case of tensile loading, LITS was not considered but a Rankine-type plasticity model was used for the simulation of tensile cracking of concrete.

- **Loading of reinforcement:** Due to the smaller stress built-up obtained from the proposed differential formulation, the stresses in the reinforcement are reduced, leading to less plastic deformations and, thus, an increase of the integrity of the tunnel-support structure.

- Reduction of bending moment: Originating from the reduced stress built-up obtained from the differential formulation, the maximum bending moment was reduced, indicating a higher structural safety of the tunnel-support structure when subjected to fire.

The presented possibility of improved modeling of the material behavior of concrete under fire loading, enabling a more realistic prediction of the thermally-induced stress build-up within the concrete lining, provides the proper basis for a realistic structural safety assessment and design.

B.7 Appendix

B.7.1 Effective prescribed strains in two-phase materials

According to [42], the effective strain E_{eff} is related to the prescribed strain $\bar{\varepsilon}$ in the material phases as:

$$K_{eff} E_{eff} = \langle A : K : \bar{\varepsilon} \rangle_V . \tag{B.19}$$

Considering a two-phase material with matrix m and inclusion i, with

$$\bar{\varepsilon} = \bar{\varepsilon}_m \text{ in } V_m , \quad \bar{\varepsilon} = \bar{\varepsilon}_i \text{ in } V_i , \tag{B.20}$$

$\bar{\varepsilon}_i$ may be substituted by

$$\bar{\varepsilon}_i = \bar{\varepsilon}_m + \Delta \bar{\varepsilon}_i . \tag{B.21}$$

Rewriting Equation (B.19) and considering Equation (B.21) gives

$$K_{eff}E_{eff} = \langle A:K\rangle_V \bar{\varepsilon}_m + f_i \langle A:K\rangle_{V_i}\Delta\bar{\varepsilon}_i. \qquad (B.22)$$

Considering $K_{eff} = \langle A:K\rangle_V$ in Equation (B.22), one gets

$$E_{eff} = \bar{\varepsilon}_m + f_i \langle A\rangle_{V_i} \frac{K_i}{K_{eff}}(\bar{\varepsilon}_i - \bar{\varepsilon}_m), \qquad (B.23)$$

where $\langle A\rangle_{V_i}$ is given in [42].

Publication C

Underground concrete frame structures subjected to fire loading: Part I – large-scale fire tests [68]

Authored by Thomas Ring, Matthias Zeiml and Roman Lackner
Published in Engineering Structures, 2012

Fire events in enclosures (such as tunnels) can have severe influence on the safety of support structures, characterized by severe fire loading as regards both the speed of temperature rise and maximum temperatures. This paper contains results of real-scale fire tests, comprising temperatures within the tested concrete cross-section, deformations and rotations at various locations as well as the extent of spalling. Monitoring of cracks during and after fire loading completes the experimentally-obtained data. Based on the collected data, validation of numerical tools for the assessment of the safety of underground support structures, as presented in Part II of this paper, becomes possible.

C.1 Introduction - Motivation

Tunnel-fire events in history have shown the severe influence of fire on concrete structures, resulting in a significant reduction of the load-carrying capacity and, thus, the safety of support structures. For assessment of the safety of concrete structures subjected to fire, the nonlinear temperature distribution within the concrete cross-section resulting from fire loading is conventionally converted into a linear temperature distribution [78]. The so-obtained linear temperature distribution can be considered easily in state-of-the-art design, employing linear-elastic analysis tools. This mode of analysis, however, has some drawbacks. E.g., stress re-distributions cannot be considered and, hence, plastic regions cannot be detected. Additionally, the deformations obtained from linear-elastic analyses using linear temperature distributions significantly underestimate the actual extent of deformations during fire loading. As a remedy, the use of finite element (FE) tools, taking non-linear temperature distributions as well as inelastic material behavior into account, was proposed in the open literature (see, e.g., [21, 71, 81]). Departing from the analysis tool presented in [71], a non-linear FE model (using layered finite elements) is currently developed within a research initiative dealing with the safety of underground structures under fire loading. Hereby, the non-linear (temperature-dependent) behavior of concrete as well as spalling of near-surface concrete layers can be considered. Moreover, non-linearities at the structural level such as the development of plastic regions/hinges as well as subsequent stress/force redistribution are taken into account.

For the proper validation of such analysis tools, experiments are indispensable. Over the last years, fire experiments focused on the material behavior, i.e. the fire performance of concrete. Hereby, concrete slabs were subjected to pre-specified fire-loading scenarios (see, e.g., [10, 26, 32, 39] and Figure C.1). As regards the investigation of the large-scale performance of tunnels, the bulk

Figure C.1: Characterization of high-temperature behavior of concrete by means of fire experiments on slabs: specimen (a) during and (b) right after the test [66]

of experiments focused on investigation of the temperature rise at the tunnel ceiling, the fire spread, the temperature evolution at certain points and/or the heat-release rate (see, e.g., [27, 28, 45, 46]). The structural performance of tunnel support structures, on the other hand, was rarely examined. Only a few tests were reported in the literature, dealing with circular cross-sections [51], disregarding any re-distribution of loading within the tunnel cross-section.

In Part I of this paper, the experimental data collected from frame structures (representing a rectangular tunnel cross-section) subjected to fire loading within the afore-mentioned research initiative are presented. In Section C.2, the geometric properties of the large-scale specimens and information on the loading (thermal, mechanical) are given. Moreover, the conducted measurements during testing (temperatures, deformations, rotations, acoustic measurements) and employed materials are described. In Section C.3, the obtained results of two concrete frames are presented, providing the evolution of deformation and temperature distributions within the cross-section as well as the extent of spalling and the development of cracks within the specimens. Concluding remarks are given in Section C.4, closing Part I of this paper.

In Part II of this paper, validation of an analysis tool is illustrated by the re-analysis of the outlined experiments, considering the loading conditions and the experimentally-obtained material parameters for concrete as input.

C.2 Experimental setup

C.2.1 Geometric properties and loading

Prior to casting of the frame, a concrete sub-base with an average thickness of 10 cm was built (see Figure C.2(a)), providing a proper support for the frame structure. Figure C.2(b) shows the geometric dimensions of the frame. In contrast to the experiments reported in the literature [51], the shape of the frame was chosen in order to represent a rectangular tunnel. Hereby, only the upper triangle of the cross-section was considered. The overall lining thickness was 40 cm. The frame had a length of 6 m and a height of 3 m. In axial direction of the tunnel, the frame had a width of 2 m.

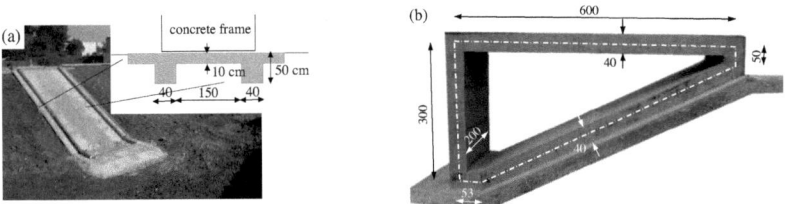

Figure C.2: Geometric dimension of (a) sub-base and (b) frame

The concrete frame (see Figure C.3(a)) was designed according to building regulations [19]. Hereby, the concrete cover at the inner (fire-exposed) surface was set to 6 cm while the concrete cover at the outer surface of the frame was set to 3 cm. Figure C.3(b) contains detailed information on the circumferential reinforcement. The reinforcement in longitudinal direction was uniform over the whole cross-section, with Ø 12/15 cm (7.54 cm^2/m) in the whole structure.

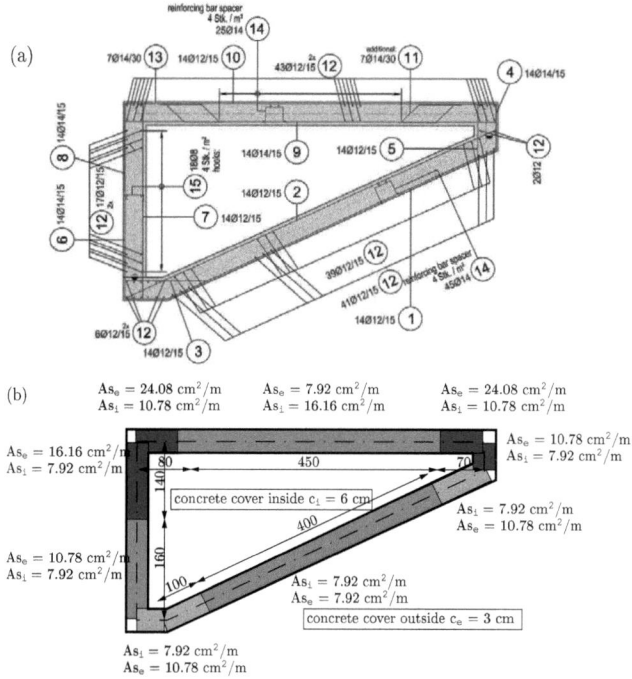

Figure C.3: Reinforcement: (a) Layout and (b) values of circumferential reinforcement (indices i and e refer to inner and outer layer, respectively)

In order to simulate the conditions of a real tunnel cross-section, an earth overburden of 170 cm was represented by steel weights of approximately 390 kN (see Figure C.4(a)). The steel weights were placed close to mid-span of the frame on three bedding points (see Figure C.4(b)).

The applied fire load is shown in Figure C.5. The maximum fire room temperature of 1200 °C was to be reached after 9 min of fire duration and thereafter held constant. The fire load was applied to the ceiling and side wall of the frame. The base plate was protected by means of insulation and additional fire-resistant insulation cover. The heating chamber was closed by mounting two removable walls to the sides of the frame. In order to reach the temper-

 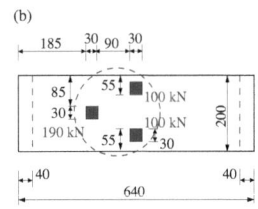

Figure C.4: Mechanical loading: (a) Steel weights (390 kN); (b) loading points on the ceiling of the frame

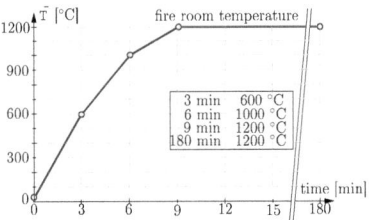

Figure C.5: Applied fire load

ature history given in Figure C.5, two oil burners with 2 MW of heat output each were used (see Figure C.6(a)). The two burners were placed on one side of the frame, whereas two chimneys were installed on the other side in order to remove the exhaust air from the heating chamber (Figure C.6(b)).

Figure C.6: Temperature loading: (a) oil burners; (b) chimneys for exhaust air

C.2.2 Material

Table C.1 contains information on the employed concrete mix, fulfilling the requirements for a concrete C30/37 [19]. Prior and during the large-scale fire

Table C.1: Mix design of employed concrete

cement CEM I [kg/m^3]	290
additive (fly ash) [kg/m^3]	50
water [kg/m^3]	185
polypropylene (PP) fibers [kg/m^3]	0
siliceous aggregates [kg/m^3]	1859
aggregate mineralogy:	
quartz [mass-%]	68
feldspar [mass-%]	21
carbonate [mass-%]	11
water/cement-ratio [–]	0.64
water/binder-ratio[†] [–]	0.56
initial density [kg/m^3]	2384
slump [mm]	430
air content [%]	1.0

[†] The amount of additives is weighted by 0.8 [53].

experiments, tests were conducted on reference samples (produced together with the frames) in order to obtain material parameters such as compressive strength, elastic modulus, thermal strain, permeability, and water content (see Figure C.7).

Figure C.7: Reference samples

Figure C.8 shows the experimental results for the compressive strength (average over three cylindrical samples, with a diameter of 100 mm and a height of

200 mm). For the compressive strength, the samples were pre-heated and the temperature was kept constant for 12 h at the testing temperature, directly followed by the compression test. Up to a temperature of approximately 300 °C, the compressive strength is hardly reduced. The main degradation of strength is obtained in the temperature range between 300 and 700 °C. For the determi-

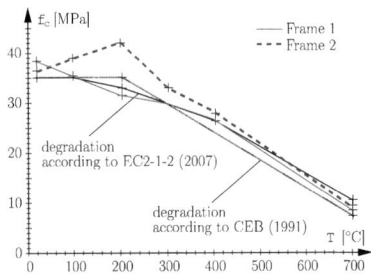

Figure C.8: Experimentally-obtained evolution of compressive strength f_c as a function of temperature compared to relations given in [11, 20] (considering $f_{c,0} = 35$ MPa)

nation of Young's modulus the samples were heated and simultaneously tested in the very same testing device (see [70] for details). The effect of temperature on Young's modulus of concrete is shown in Figure C.9. A continuous loss of stiffness is observed which is less pronounced between 200 and 300 °C.

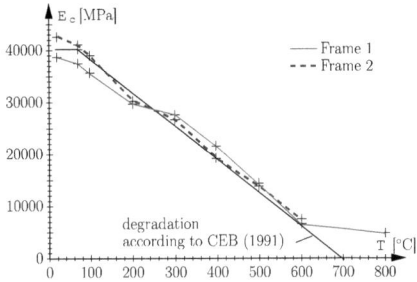

Figure C.9: Experimentally-obtained evolution of Young's modulus E_c as a function of temperature compared to relation given in [11] (considering $E_{c,0} = 40000$ MPa)

Additionally, the free thermal expansion was tested (see Figure C.10) and compared to the free thermal expansion given in [20] for siliceous aggregates, indicating higher expansion for the investigated concrete mix, mainly due to higher expansion between 20 and 200 °C. As the behavior of free thermal strain of concrete is mainly driven by the used siliceous aggregates, a plateau in the thermal expansion develops after the point of quartz transition (573 °C) followed by marginal changes in the thermal strains [33].

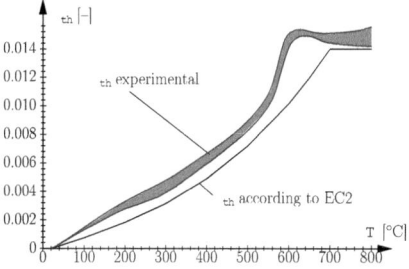

Figure C.10: Experimentally-obtained evolution of free thermal expansion as a function of temperature compared to evolution given in [20] for siliceous aggregates

For determination of the permeability of the employed concrete mix, specimens were stored for 12 h at different target temperatures (20, 80, 140, 200, and 400 °C) prior to testing. After cooling to room temperature, the residual permeability was determined using the testing device described in [24, 80]. According to Figure C.11, the permeability increases with increasing temperature, which is explained by thermal degradation of the cement paste as well as continuous micro-cracking due to strain incompatibilities between aggregates and cement paste.

The moisture content within the frames was measured before testing by drilling holes into the specimen and collecting the drill dust. For determination of the moisture content, this dust was dried at 105 °C for 24 h. The so-obtained results are presented in Figure C.12. Additionally, the dust was further dried at 500 °C for another 24 h, giving access to the chemically-bound water content, being 2 to 2.5 mass-%.

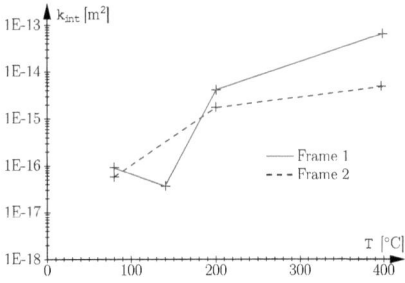

Figure C.11: Experimentally-obtained evolution of intrinsic permeability

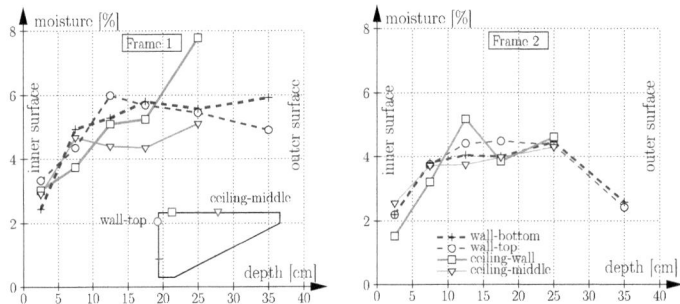

Figure C.12: Moisture content (by mass) within the frames determined by drying at 105 °C for 24 h

C.2.3 Measurements

In order to apply the prescribed temperature loading, the two oil burners described before were regulated based on the data obtained with seven sensors placed inside the combustion chamber (see Figure C.13(a)). In addition to these sensors, temperature sensors were placed within the cross-section of the concrete frame (see Figure C.13(b)). At each location specified in Figure C.13(b), the temperature history was recorded at different depths of the cross-section (0.3, 1, 2, 4, 6, 8, 10, 20, and 30 cm from the heated surface). For this purpose, precast sensor members (see Figure C.14) were placed within the frame before concreting.

Moreover, the temperature of the reinforcement (inner layer and outer layer)

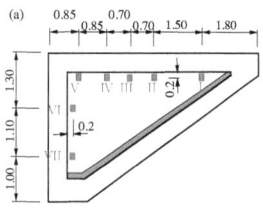 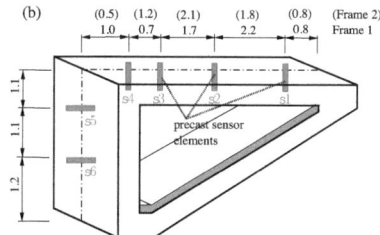

Figure C.13: Location of temperature sensors within (a) the combustion chamber and (b) the concrete frame

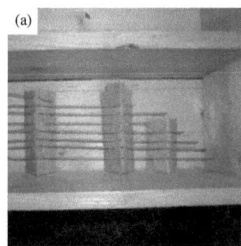 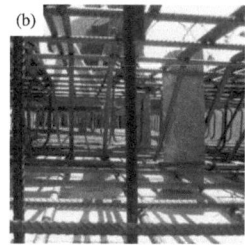

Figure C.14: Temperature measurements: (a) Mold for producing precast element with sensors at different depths (0.3, 1, 2, 4, 6, 8, 10, 20, and 30 cm); (b) precast sensor element placed within the frame before concreting

was recorded at various locations. For checking purposes, the reinforcement temperature was also recorded in the (thermally-insulated) base plate.

Deformations of the frame (in X/Y/Z-direction) were recorded at 31 locations distributed over the ceiling and side wall of the frame (see Figure C.15). Measurements were conducted automatically by two theodolites, every measurement point was monitored at a time interval of approximately 2 minutes. The mirrors mounted to the frames (see Figure C.15), measuring the displacement at a distance of 19.7 cm to the frame surface, were protected against hot exhaust air by sufficient insulation and an additional air-cooling system. In addition to the deformation measurements, rotations were recorded at selected locations by inclinometers (see Figure C.15).

Beside temperature and deformation/rotation measurements, the intensity

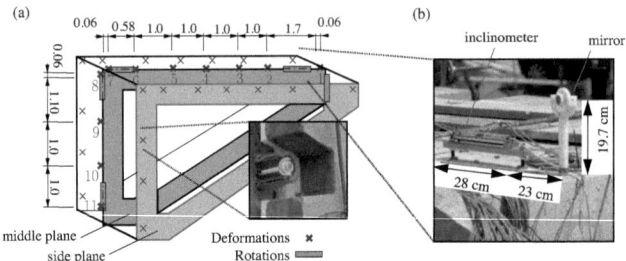

Figure C.15: Deformation measurements: (a) positions of mirrors (for deformation measurements) and inclinometers (for rotation measurements); (b) Detailed position of inclinometer and mirror

of spalling was recorded by means of acoustic measurements. Moreover, the final spalling depth as well as location and width of cracks were measured after each experiment.

C.3 Experimental results and discussion

C.3.1 Temperature measurements

The temperature evolution in the combustion chamber, which was measured at seven locations (see Figure C.13(a)), is presented in Figures C.16, indicating that the prescribed maximum temperature (with 1200 °C after 9 min of fire loading) was reached after about 35 min of fire loading. This delay in heating is explained by the performance of the employed oil burners and the occurrence of spalling, requiring heating of colder concrete layers exposed by spalling. After most spalling events had taken place within the first 30 min of heating, the prescribed fire curve was reached after this time period. Sensors I and VII show lower temperatures in comparison to the other temperature sensors in the combustion chamber, which is explained by spalling material eventually covering the temperature sensors in this regions.

The temperature evolution within the concrete cross-section at different depths

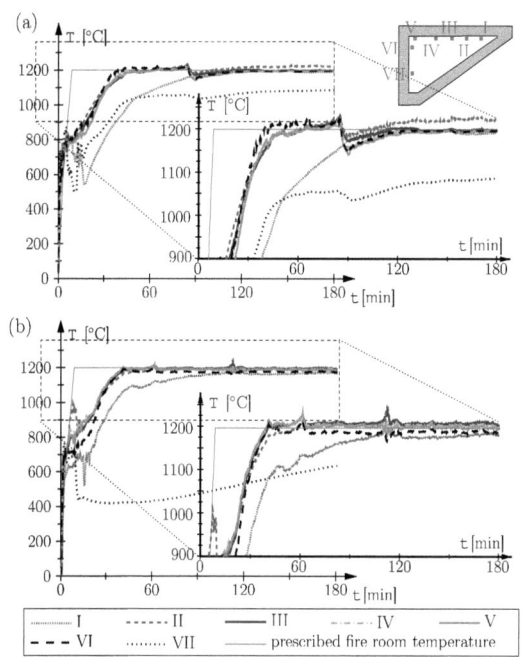

Figure C.16: Monitored evolution of combustion-chamber temperature (a) for Frame 1 and (b) Frame 2

of the cross-section (0.3, 1, 2, 4, 6, 8, 10, 20, and 30 cm) is shown in Figures C.17 and C.18. The temperature evolutions (see, e.g., Figure C.17) more specifically the temperature rise in the ceiling (Sensors 1 to 4) and in the side wall (Sensors 5 and 6) give access to the extent of spalling. As soon as the measured temperature in a certain depth (0.3, 1 cm, ...) shows temperatures $\geq 1200\,°C$, spalling of the concrete cover to the depth of the respective temperature sensor can be assumed, giving (e.g. for Sensors 1 and 6) a final spalling depth of approximately 6 cm. The temperature sensor at a depth of 30 cm exhibits a maximum temperature of approximately 105 °C (the observed temperature plateau is caused by vaporization of water within the cross-section). Sensors 5 and 6 show temperatures higher than 1200 °C, which is explained by the ar-

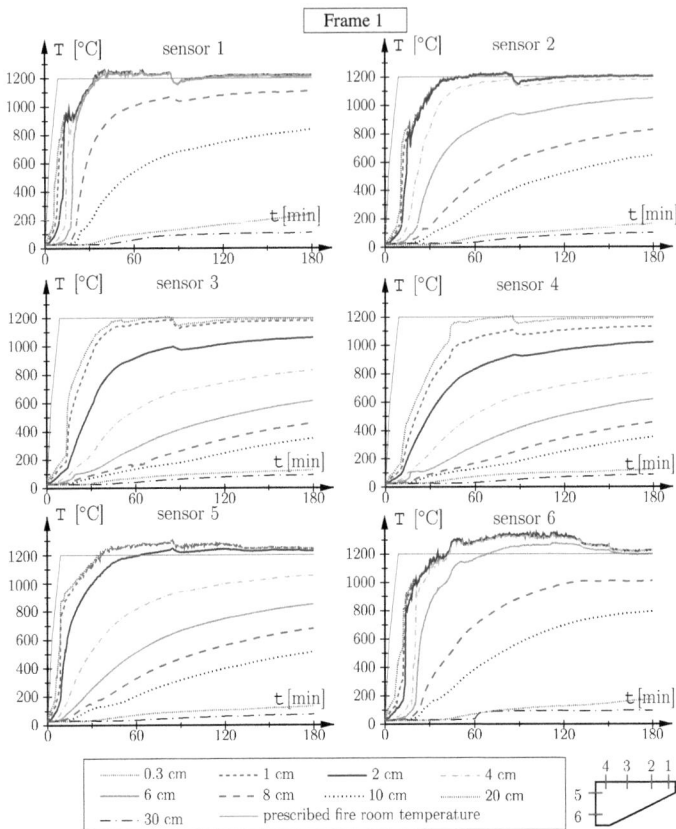

Figure C.17: Monitored evolution of concrete temperature at different depths for Frame 1

rangement of the oil-burners, leading to higher temperature loading of the side wall in order to provide a uniform temperature loading all over the combustion chamber.

In Figure C.18, the monitored temperature evolutions for Frame 2 are presented. The arrangement of the oil-burners was slightly modified, resulting in lower temperatures at Sensors 4 to 6. Similarly to Frame 1, the final spalling depth reached 6 cm (e.g. for Sensor 2). The remarkable increase in temperature at Sensor 6 was caused by a local spalling event, which was also recognized

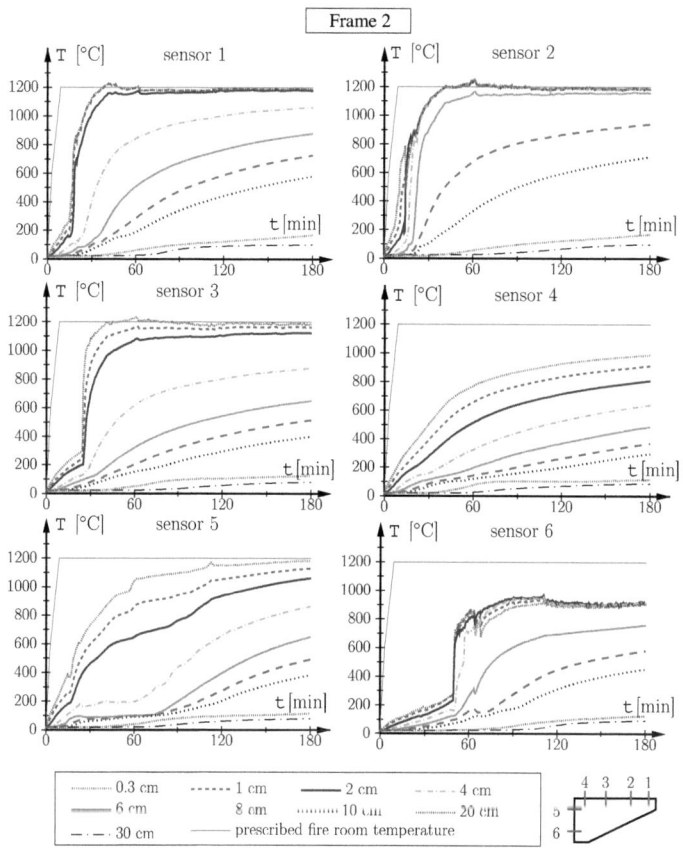

Figure C.18: Monitored evolution of concrete temperature at different depths for Frame 2

acoustically.

Figure C.19 contains the evolution of the temperature at the reinforcement, with higher reinforcement temperatures for Frame 1 in comparison to Frame 2. This corresponds to the previously presented temperature evolutions within the cross-section. Only the inner reinforcement layers of Frame 1 at Sensors 1 and 2 are directly exposed to fire (temperature reaches 1200 °C). The outer reinforcement layer shows no significant temperature rise up to 60 min of fire loading. Later on, the temperature slightly increases and does not exceed 100 °C.

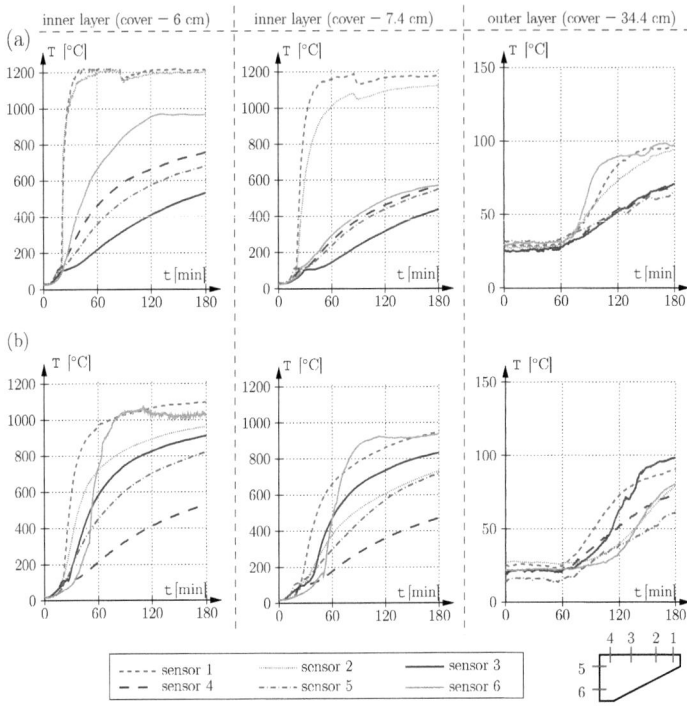

Figure C.19: Monitored evolution of temperature at reinforcement layers for (a) Frame 1 and (b) Frame 2

C.3.2 Deformation measurements

Prior to application of the mechanical load, the deformation readings of all points were set to zero. The deformation history of the frames during application of the mechanical loading is presented in Figure C.20, showing a maximum vertical deformation of approximately 4 mm at Point 4 (close to the mid-span of the frame ceiling). The results for both frames are very similar. The vertical deformation at Point 11 indicates an immediate settlement of the frame (1 to 2 mm) during application of the mechanical load.

The time span between application of the mechanical load and the fire experiment (approximately two weeks) gives access to the time-dependent behavior

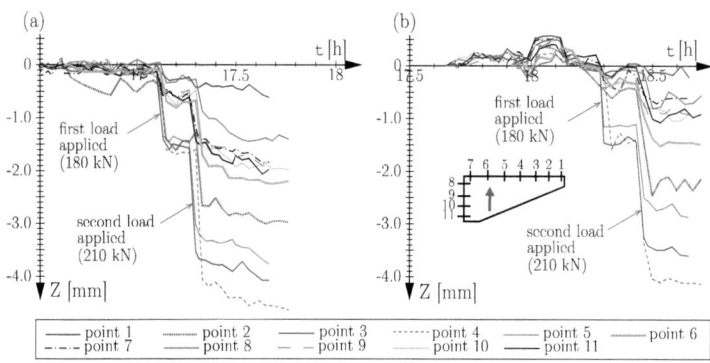

Figure C.20: Monitored vertical deformations in the symmetry plane of the frames due to mechanical loading approximately 2 weeks prior to testing (no fire load) (a) for Frame 1 and (b) Frame 2

of the frames (see Tables C.2 and C.3) as well as to the settlement of the whole

Table C.2: Deformations at symmetry plane for Frame 1

Pt.	absolute coordinates before mechanical loading (reference) – 11/05/2010			Δ to reference after mechanical loading – 11/05/2010			Δ to reference before fire exp. – 09/06/2010		
	X [m]	Y [m]	Z [m]	ΔX [mm]	ΔY [mm]	ΔZ [mm]	ΔX [mm]	ΔY [mm]	ΔZ [mm]
1	1021.852	982.998	200.237	0.0	1.2	-0.6	2.0	4.0	-2.3
2	1021.717	984.624	200.239	-0.7	-1.3	-3.0	8.5	7.3	-6.9
3	1021.654	985.621	200.238	0.0	1.4	-4.1	0.6	4.5	-9.8
4	1021.588	986.628	200.245	0.3	0.9	-4.7	0.3	2.2	-10.4
5	1021.520	987.617	200.241	0.5	0.5	-3.8	0.0	2.1	-9.7
6	1021.455	000.646	200.238	1.0	1.6	-2.2	2.4	3.5	-8.1
7	1021.407	989.270	200.235	0.0	0.9	-1.9	-0.4	4.2	-7.0
8	1021.401	989.539	199.964	0.1	0.9	-1.4	0.1	2.4	-6.4
9	1021.397	989.534	198.858	-0.1	1.3	-1.9	-0.3	2.7	-6.7
10	1021.394	989.532	197.860	0.2	0.6	-2.0	-0.2	1.3	-7.1
11	1021.392	989.519	196.833	-0.3	0.1	-2.0	0.0	1.0	-5.9

frame when considering the deformations at Points 1 and 11. E.g. for Points 3 to 5 (mid-span of frame ceiling), these deformation changes up to 6 mm (in self weight (Z–) direction).

In the course of fire loading, the symmetry plane experienced the largest deformations (up to 70 mm) in the vertical (Z–) direction (see Figure C.21). Hereby, Frame 1 shows larger deformations than Frame 2, which is explained by the higher fire loading (compare Figures C.17 and C.18). The deformations

Table C.3: Deformations at symmetry plane for Frame 2

Pt.	absolute coordinates before mechanical loading (reference) – 23/06/2010			Δ to reference after mechanical loading – 23/05/2010			Δ to reference before fire exp. – 07/07/2010		
	X [m]	Y [m]	Z [m]	ΔX [mm]	ΔY [mm]	ΔZ [mm]	ΔX [mm]	ΔY [mm]	ΔZ [mm]
1	1022.149	979.107	200.234	-0.3	-1.6	-0.2	2.1	-0.8	-1.4
2	1022.267	977.462	200.232	0.7	-0.4	-2.3	-1.5	-5.2	-5.4
3	1022.345	976.470	200.227	0.3	-1.0	-3.6	2.5	-2.8	-7.7
4	1022.388	975.482	200.227	0.5	0.1	-4.2	-5.5	-1.2	-8.9
5	1022.508	974.479	200.227	0.4	-0.2	-2.9	3.1	-0.2	-6.5
6	1022.577	973.497	200.233	0.5	0.4	-1.5	-0.5	1.7	-4.7
7	1022.629	972.849	200.231	0.4	-0.3	-0.6	0.9	0.5	-2.2
8	1022.622	972.589	199.963	0.7	0.2	-0.9	-1.7	0.3	-2.6
9	1022.632	972.591	198.822	0.3	-1.0	-0.9	4.7	-1.5	-3.2
10	1022.639	972.590	197.826	0.5	-0.7	-0.9	6.8	-1.2	0.5
11	1022.649	972.607	196.990	0.3	0.7	-1.0	-1.0	-0.2	-2.4

in horizontal (X-) direction mainly result from the thermal expansion of the frame. After the fire experiment, the mechanical load was removed from the frames, giving a sudden decrease in the monitored deformations (visible in Y- and Z-direction, see Figures C.21 and C.22).

In addition to the symmetry plane, the deformations of the front plane are presented in Figure C.22. Comparing Figures C.21 and C.22 gives access to the bulge of the frame in tangential direction (1 m distance between symmetry and front plane, see Figure C.2). The largest value for this bulge is monitored at the mid-span of the ceiling and reaches approximately 20 mm after 180 min. Finally, the deformations in X-direction provide access to the expansion of the frame in tangential direction.

Figure C.23 shows the rotations monitored during application of the mechanical load. Two jumps are visible, referring to the two steel weights (180 + 210 = 390 kN). The rotations at the two frame corners (Corner 1 = Points 1 and 2, Corner 2 = Points 3 and 4), show a maximum value of approximately 1 mm/m. The rotations of the frame at the bottom of the wall, on the other hand, remain rather small (see Point 5 in Figure C.23). Looking at the rotations

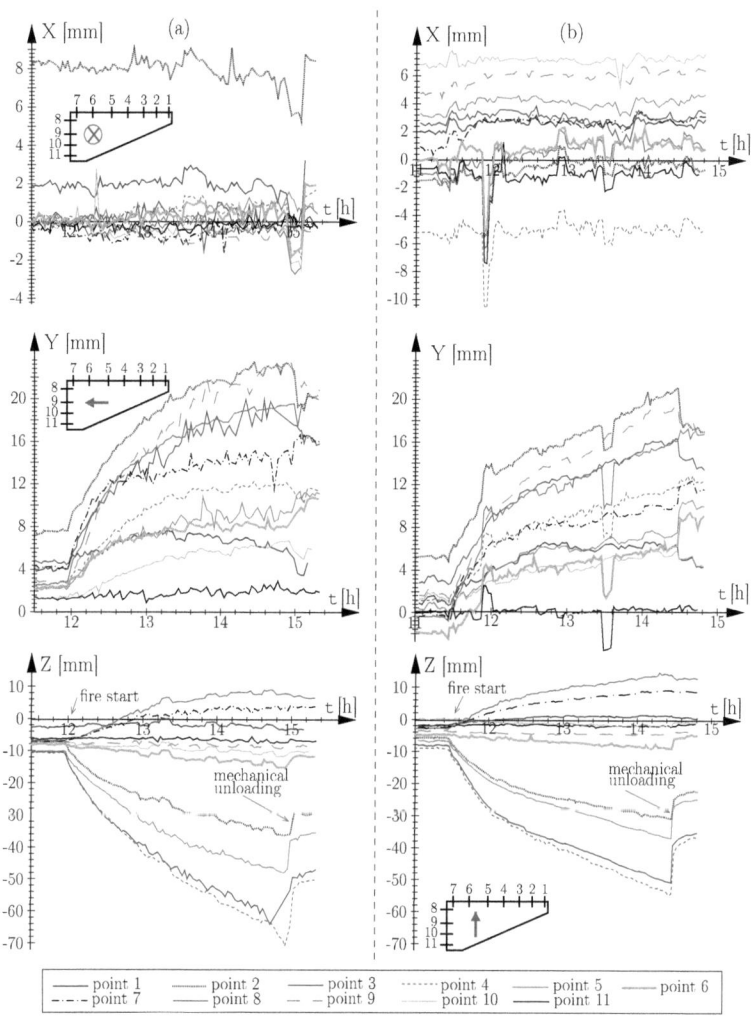

Figure C.21: Monitored evolution of deformations (X/Y/Z-direction) during fire experiment at points of the symmetry plane (a) for Frame 1 and (b) Frame 2

during fire loading (see Figure C.24), one can see that the rotation at Point 5 hardly changes, while the rotations at the two frame corners show a significant increase, with the corner at the ceiling/wall connection (Corner 2) showing the largest rotations (up to 28 mm/m, see Figure C.24). In general it can be seen

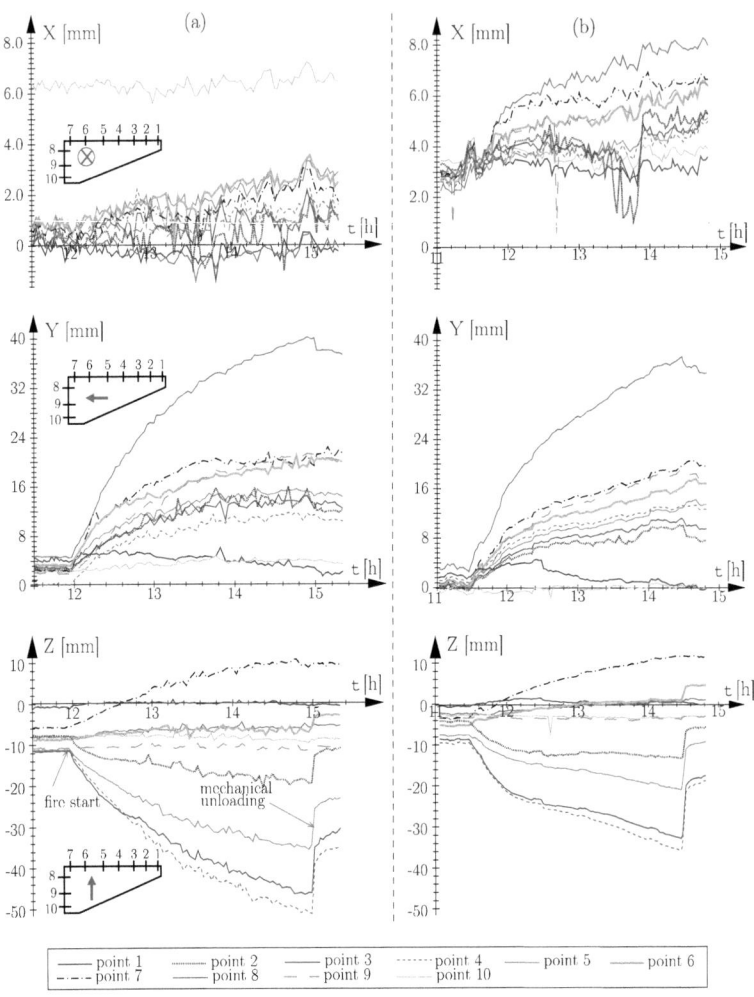

Figure C.22: Monitored evolution of deformations (X/Y/Z-direction) during fire experiment at points of the front plane (a) for Frame 1 and (b) Frame 2

that rotations at the two respective points of the two monitored frame corners are similar to each other, indicating that no cracks opened at the frame corners. After the fire experiment, removal of the mechanical load resulted in a sudden change in the evolution of rotations.

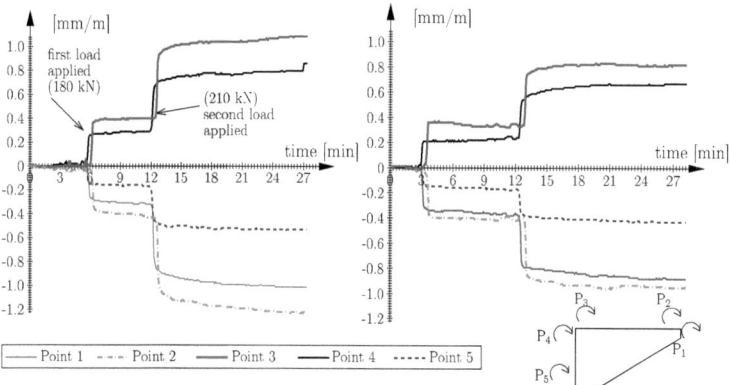

Figure C.23: Monitored evolution of rotations at five points prior to fire loading (a) for Frame 1 and (b) Frame 2

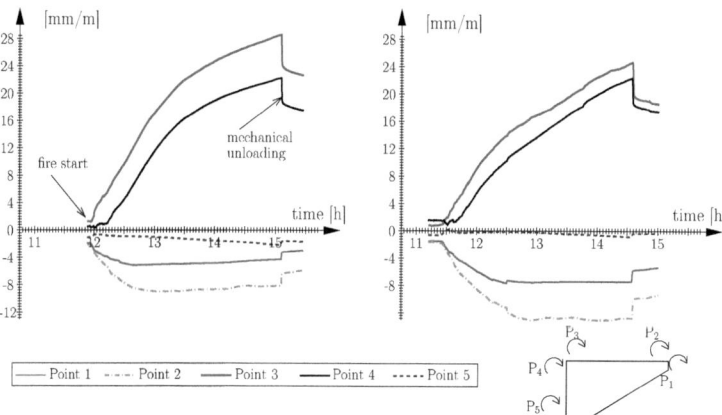

Figure C.24: Monitored evolution of rotations at five points during fire loading (a) for Frame 1 and (b) Frame 2

C.3.3 Spalling and cracks

Spalling was monitored by means of acceleration sensors mounted to the frames, recording the impact sound. Figure C.25 shows the sound level for different frequencies (between 100 and 10800 Hz) for the first 30 min of fire loading. Peaks greater than 100 dB are projected onto the base plane (i.e., sound level = 0 dB). Accordingly, the main part of spalling is found within 3 to 25 min of testing (which is in agreement with other experiments, see, e.g., [39]). Afterwards, only random spalling events were observed. Summing up the single events with a

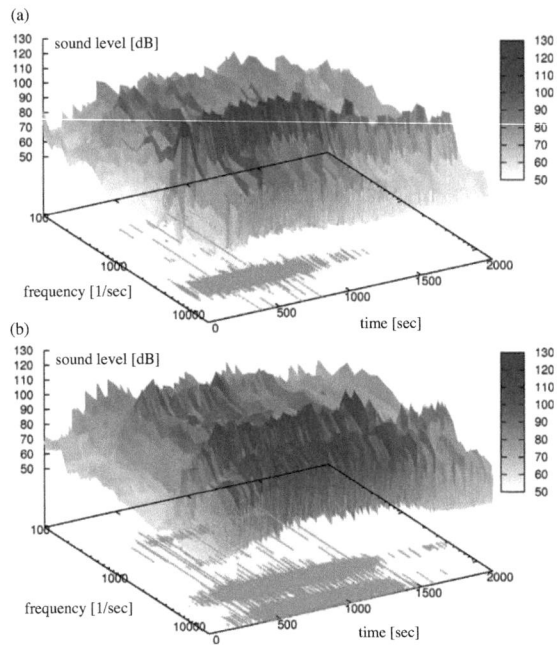

Figure C.25: Monitored evolution of sound level at different frequencies (a) for Frame 1 and (b) Frame 2

sound level larger than 100 dB leads to the evolution of the progress of spalling within the frame (see Figure C.26). The observed delay in spalling for Frame 2 is explained by the higher thermal load at the beginning of fire loading for Frame 1, leading to a faster development of spalling.

In order to assess the extent of spalling, the remaining cross-sectional thickness was measured after the fire experiment, giving access to the final spalling depth as shown in Figures C.27 and C.28 for Frame 1. Hereby, spalling depths up to 6.5 cm were observed, partially exposing the inner reinforcement layers. Figure C.27 shows the final spalling depth at the inner surface of the side wall

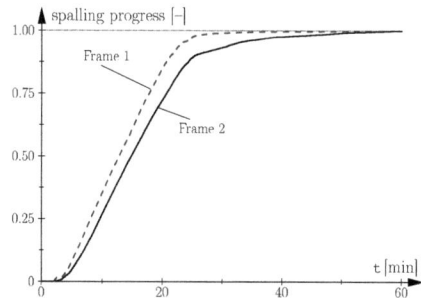

Figure C.26: History of extent of spalling for Frames 1 and 2

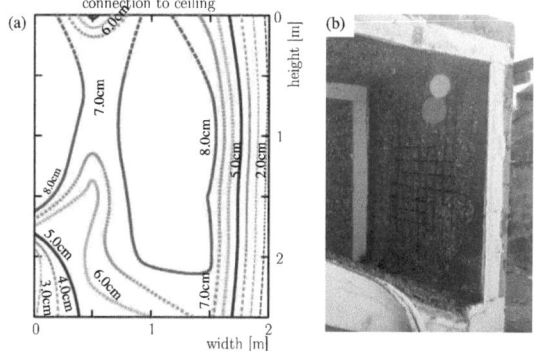

Figure C.27: Final spalling at the side wall for Frame 1: (a) spalling depth and (b) photo right after the experiment

(where the highest temperatures were measured), indicating fire exposure of the inner reinforcement layers caused by spalling. The ceiling, on the other hand, shows only moderate spalling (see Figure C.28) with only local (near Sensor 6) exposure of the reinforcement. This illustrates the influence of mechanical loading: the wall (being under compressive loading) exhibits more spalling compared to the ceiling (with the fire-loaded surface under flexural tension).

For Frame 2, the final spalling depth is presented in Figures C.29 and C.30, showing a similar behavior as observed for Frame 1. Similarly to Frame 1, the largest amount of spalling is observed at the side wall (compare Figures C.27(a) and C.29(a)). The ceiling of Frame 2 (compare Figures C.28(a) and C.30(a))

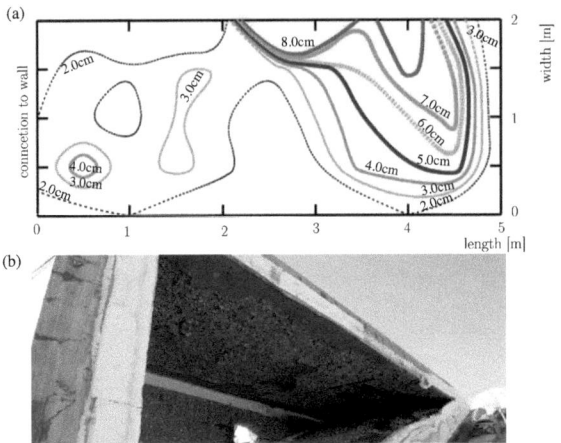

Figure C.28: Final spalling at the ceiling for Frame 1: (a) spalling depth and (b) photo right after the experiment

experiences a slightly smaller extent of spalling.

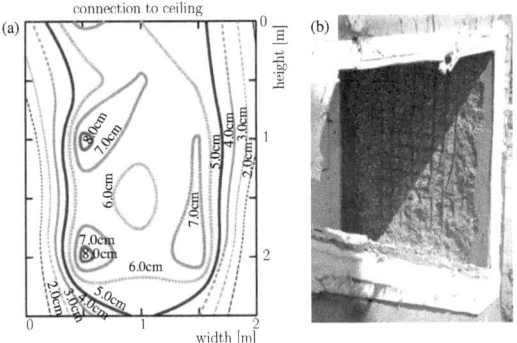

Figure C.29: Final spalling at the side wall for Frame 2: (a) spalling depth and (b) photo right after the experiment

Two days after the experiment, the crack width of all visible cracks was measured (see Figures C.31 and C.32). The so-obtained residual cracks (after removal of mechanical loading and cooling of the frames) were found to be concentrated near the frame corners, having experienced the highest bending moments, with crack widths below 1.5 mm. At the corner of ceiling and wall,

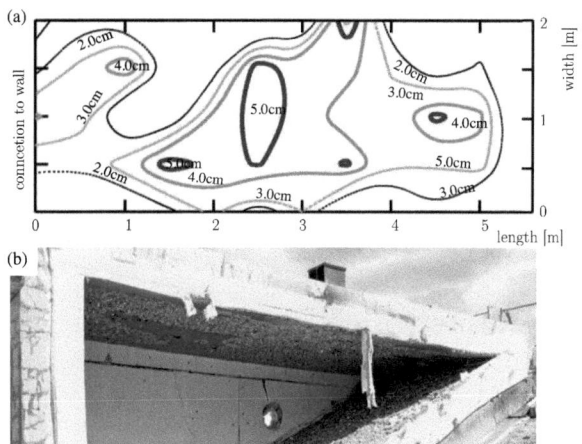

Figure C.30: Final spalling at the ceiling for Frame 2: (a) spalling depth and (b) photo right after the experiment

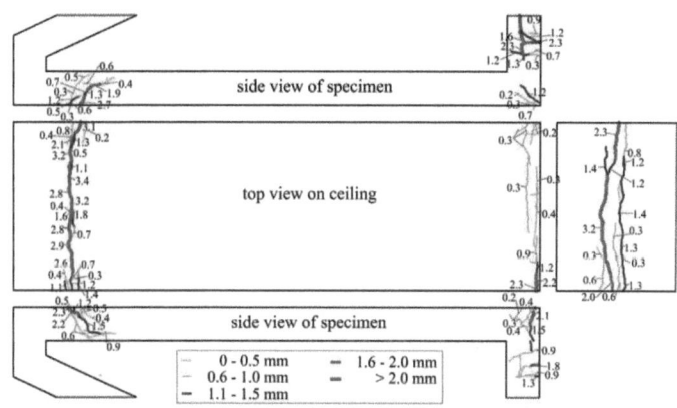

Figure C.31: Residual cracks for Frame 1 after cooling and removal of mechanical load

cracks developed below the corner, where the cross-sectional thickness was reduced due to spalling to less than 35 cm. Comparing the cracks for Frames 1 and 2, the smaller crack widths for Frame 2 agree well with the lower temperature loading observed for this frame.

Figure C.33 shows selected crack patterns. A fine grid of cracks was observed during the fire experiment (see Figure C.33(a)), which vanished (cracks closed)

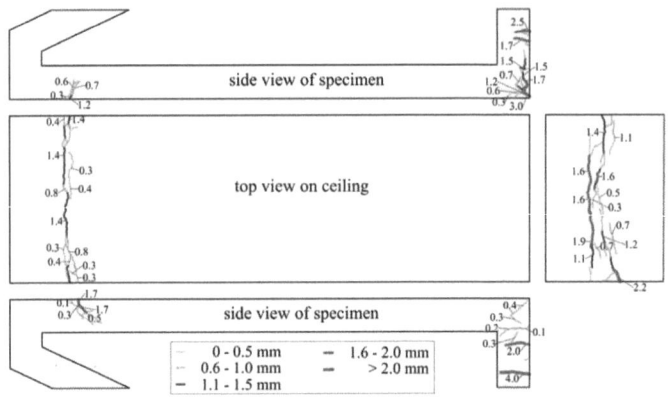

Figure C.32: Residual cracks for Frame 2 after cooling and removal of mechanical load

after removal of the steel weights and is therefore not included in Figures C.31 and C.32. Figure C.33(b) shows a macro-crack caused by combined mechanical and thermal loading in the course of the fire experiment. This crack seems to be triggered by the reinforcement, which was bent upwards 45° according to the reinforcement design (see Figure C.3).

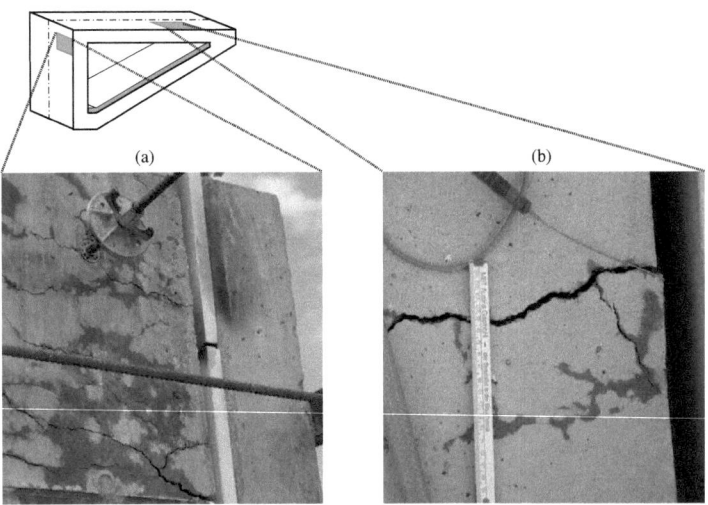

Figure C.33: Cracks during fire experiment: (a) Fine net of smaller cracks, (b) macro-crack

C.4 Concluding remarks

In this paper, the setup and results of large-scale fire experiments on underground concrete frames (representing a rectangular tunnel) were presented. The monitored experimental data is considered to provide a proper basis for validation of analysis tools, such as:

- Hydro-thermo-chemical programs (see, e.g., [81]), providing access to temperature evolutions within the cross-section as a function of the initial state of concrete (such as e.g. humidity, porosity, etc.) and the applied fire loading.

- Hydro-thermo-chemical programs (see, e.g., [82, 84]) for the analysis of the spalling process.

- Structural analysis tools for the simulation of the structural response (see, e.g., [71, 83]).

As regards the latter, a structural safety assessment tool (using layered finite elements, as earlier presented in [71]) is refined and presented in Part II of this paper, using the collected data as experimental basis for validation.

Publication D

Underground concrete frame structures subjected to fire loading: Part II – re-analysis of large-scale fire tests [69]

Authored by Thomas Ring, Matthias Zeiml and Roman Lackner

Published in Engineering Structures, 2012

The severe effect of fire loading on the performance of concrete structures is investigated in this work. For this purpose, numerical analysis tools of different levels of complexity ranging from simulations considering linear-elastic material behavior to more sophisticated modes of modeling the strain behavior of heated concrete are employed. For validation of the numerically-obtained results, large-scale fire tests on concrete frames were performed (see [68]), allowing to compare the model response to the experimentally-determined structural behavior. The performed simulations highlight the influence of model assumptions, e.g., linear-elastic material behavior or spalling, on the model response of reinforced concrete structures subjected to fire loading.

D.1 Introduction

Numerical simulations in state-of-the-art fire design are in general based on the equivalent-temperature concept [64, 78]. Within this concept, the non-linear temperature distribution within the concrete member resulting from fire loading at the heated surface is transferred into a linear temperature distribution by means of a clamped-beam model, enforcing equivalent internal stress resultants (see Figure D.1). The equivalent temperature distribution is defined by a temperature rise T_m uniform over the cross-section and a temperature gradient ΔT over the cross-section. Both of them can be considered within existing design

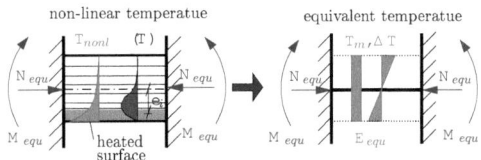

Figure D.1: Illustration of transfer of non-linear temperature distribution into equivalent temperature distribution by means of clamped-beam model [78, 39]

software using beam elements [64, 78].

Since this simplification has significant drawbacks (missing consideration of re-distribution of mechanical loading, deformations are highly underestimated by linear-elastic approach,...), more sophisticated material models, taking the nonlinear temperature distribution into account, were proposed in the past [17, 71]. Moreover, by using layered shell elements, spalling can be considered by de-activation of near-surface layers.

Departing from the analysis model presented in [71], the influence of different parameters (tensile strength, load induced thermal stains – LITS) on the structural response of the concrete frames subjected to fire is investigated in this paper. For this purpose, the experimental data presented in Part I of this paper are used for validation of the different modeling approaches. In Section D.2, the numerical model is presented, including the used FE-mesh, reinforcement scheme, boundary conditions, and loading of the investigated frame. The employed material models for concrete and steel are described in Section D.3, followed by the presentation of the obtained numerical results in Section D.4. Finally, concluding remarks are given in Section D.5.

D.2 Numerical model

D.2.1 FE-Model

In Figure D.2(a), the geometry of the investigated reinforced-concrete frame is presented. The thickness of the frame is 40 cm. In this paper, the reinforcement was considered by means of layered finite elements by converting the reinforcement bars of the respective reinforcement layer into a steel layer of equivalent thickness (see Figures D.2(b) and (c)). In contrast to continuum elements, a lower number of elements is required when using layered finite elements. Moreover, 105 layers within each finite element, allowed us to account for the non-linear temperature distribution – especially at the heated surface – within the cross-section of the concrete members.

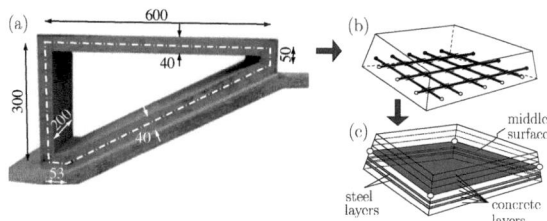

Figure D.2: (a) Dimensions (in cm) of the investigated frame [68]; layered finite element: (b) real situation and (c) consideration of reinforcement by steel layers [71]

The frame was discretized by 59 finite elements in circumferential direction and 6 finite elements in longitudinal direction, giving in total 354 shell elements (see Figure D.3). The chosen number of six finite elements in longitudinal direction allows the simulation of the bulge of the cross-section observed in the experiments [68].

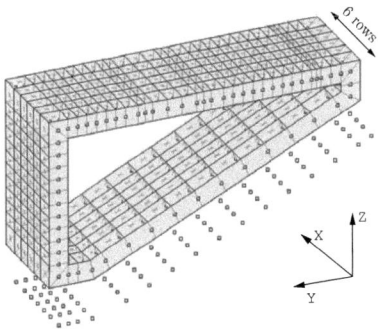

Figure D.3: Finite-element mesh of the reinforced-concrete frame

D.2.2 Boundary conditions

Within the numerical model, the frame structure was supported by means of supports at line A-A fixed in Y- and Z-direction (see Figure D.4) and at the symmetry plane in X-direction. The support by the soil at the frame base was considered by so-called gap elements (spring support only acting in compression, no support in tension) with an assumed stiffness of 30 MN/m^2 (see Figure D.4).

D.2.3 Loading

In addition to the self weight, the loading of the frame consisted of:

1. mechanical loading of 390 kN applied at three points of the ceiling according to the experimental setup (locations are shown in Figure D.5(a)) and

2. thermal loading applied at the inside surface of the ceiling and the wall. Since the bottom of the frame was insulated during the experiment (see

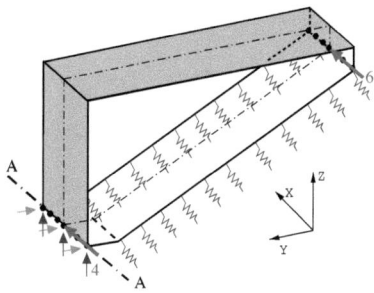

Figure D.4: Boundary conditions: Fixed supports along line A-A together with spring support at the frame base

[68] for details), no temperature loading was considered at this part of the frame. Based on the temperatures monitored during the experiment (at six locations of the concrete frame), six regions were distinguished for the

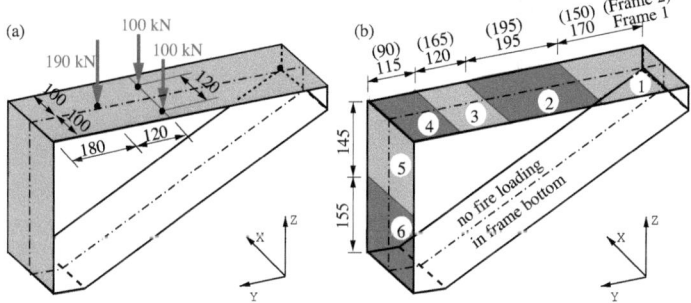

Figure D.5: Loading of frame structure: (a) mechanical loading (390 kN) and (b) thermal loading (six regions of different temperature loading were defined)

specification of the temperature load (see Figure D.5(b)). The temperature distribution within each region was set uniform. This simplification leads to an overestimation of the temperatures in the frame corner. The underlying model might be further improved by refining the temperature loading by means of additional thermal simulations.

D.2.4 Material degradation

The temperature-dependent evolution of Young's modulus and compressive strength were experimentally determined (see Table D.1 and [68] for details). Both the temperature dependence of the tensile strength of concrete (with f_{t0} =

Table D.1: Temperature-dependent evolution of Young's modulus and compressive strength of concrete (Frame 1 and 2, see [68])

Temperature [°C]	Young's modulus		Compressive strength	
	Frame 1 [MPa]	Frame 2 [MPa]	Frame 1 [MPa]	Frame 2 [MPa]
20	38660	42690	38.3	36.4
70	37380	40970	–	–
100	35510	38840	–	–
200	29740	30440	31.6	42.1
300	27570	26630	29.8	33.6
400	21360	19080	26.7	28.3
600	6430	6690	–	–
700	5620	–	8.5	9.1
1200	0	0	0	0

3 MPa at ambient room temperature) and the material properties for steel (BSt 550) are taken from [20] (see Table D.2).

Table D.2: Temperature-dependent evolution of Young's modulus and yield strength for steel (BSt 550) [20]

Temperature [°C]	Young's modulus [MPa]	Yield strength [MPa]
20	210000	550
100	210000	550
200	189000	550
300	168000	550
400	147000	550
500	126000	429
600	65100	259
800	18900	61
1200	0	0

The free thermal strain of concrete and steel was taken from [68] and [20], respectively. A comparison of the free thermal strain for concrete and steel is presented in Figure D.6.

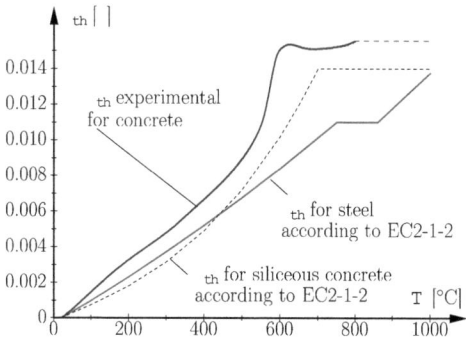

Figure D.6: Free thermal strain for concrete and steel (see [68] and [20])

D.3 Simulation parameters

D.3.1 Linear-elastic material behavior + equivalent temperature distribution

The ideal-elastic material model is used when considering the equivalent temperature loading described in Section D.1. Hereby, ideal-elastic material behavior is assumed for both concrete and steel, where the thermal degradation of Young's modulus is taken into account (see Tables D.1 and D.2). The equivalent-temperature loading for the six regions shown in Figure D.5(b) is obtained from (see [39] for details)

$$T_m = \frac{N_{equ}}{E_{equ}A} \quad \text{and} \quad \Delta T = T_i - T_e = \frac{M_{equ}}{E_{equ}I}, \tag{D.1}$$

the evolution of T_m [°C] and ΔT [K/m] is shown for Frame 1 in Figure D.7.

In Equation (D.1), A [m^2] and I [m^4] are the cross-sectional area and the moment of inertia, respectively, whereas [K^{-1}] is the thermal expansion coefficient of concrete. E_{equ} [MPa] is the equivalent Young's modulus, given by

$$E_{equ} = \sum_{i=1}^{N} \frac{E_{c,i}(T_i)A_i}{A}, \tag{D.2}$$

where $E_{c,i}(T_i)$ [MPa] and A_i [m^2] are Young's modulus and cross-sectional area of the i-th layer (with N [-] as the number of layers). The so-obtained parameters T_m, ΔT, and E_{equ} serve as input for the linear-elastic structural analysis.

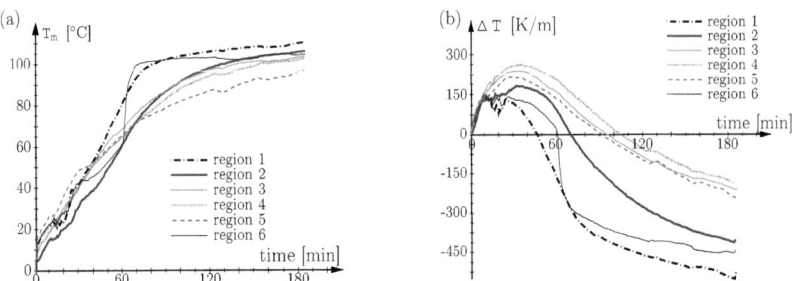

Figure D.7: Equivalent temperature loading for Frame 1 for the six domains: (a) uniform temperature increase, T_m, and (b) temperature gradient over cross-section, ΔT

D.3.2 Elasto-plastic material behavior + nonlinear temperature distribution

The in-plane behavior of concrete is described by a multi-surface plasticity model for plane stress (see Figure D.8(a) and [71] for details). The material behavior under compressive loading is described by a Drucker-Prager criterion. For the simulation of cracking, the Rankine criterion is employed. For the in-plane material model for steel, a 1D model is applied (assuming perfect bond between concrete and steel), where hardening/softening behavior of the reinforcement is considered (see [71] for details)[1].

Within the analyses based on the elasto-plastic material model for concrete and steel, the non-linear temperature distributions (see Figure D.8(b)), derived from the temperature evolutions measured during the fire experiment (see [68] for details), are considered according to the regions displayed in Figure D.7(b).

The underlying stress-strain curves for the temperature-dependent behavior

[1] The reinforcement content presented in Part 1 of this paper [68] and considered in the numerical simulation, does not account for overlapping areas of the reinforcement bars.

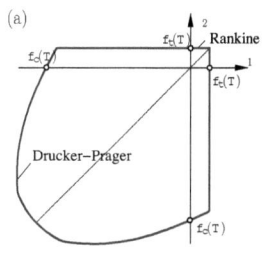
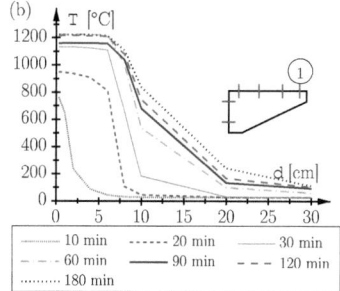

Figure D.8: (a) In-plane plasticity model for concrete; (b) temperature distributions for different time instants for Region 1 of Frame 1

of concrete under compression are shown in Figure D.9(a), highlighting the difference between the linear-elastic and the elasto-plastic model. Within the elasto-plastic model, associative hardening plasticity is considered, increasing the yield strength from the elastic limit at $0.4 f_c(T)$ to the compressive strength at the respective temperature (for further details on, e.g., hardening rule, see [71]). The temperature-dependent model for concrete under tensile failure is shown in Figure D.9(b), exhibiting linear-elastic behavior up to the tensile strength followed by an ideal-plastic material behavior. In more refined modeling attempts softening behavior including the effect of tension-stiffening [40, 41] may be considered.

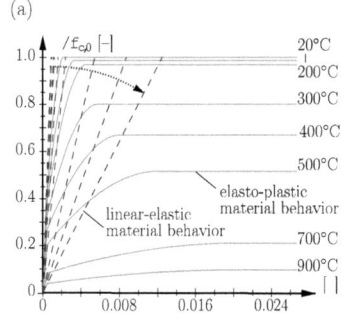
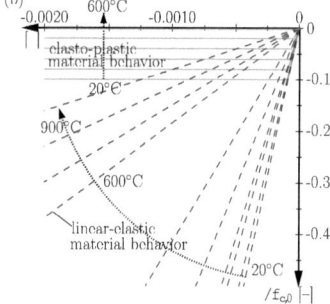

Figure D.9: Stress-strain relationship of linear-elastic and elasto-plastic material behavior for (a) compressive and (b) tensile loading of concrete

D.4 Numerical results

In this paper, different simulations were conducted, allowing us to assess the effect of, e.g. spalling, load induced thermal strains (LITS) etc. on the overall structural response:

- Model 1: Linear-elastic material behavior + equivalent temperature model (state-of-the-art)
- Model 2: Elasto-plastic material behavior + non-linear temperature
- Model 3: Model 2 + consideration of spalling
- Model 4: Model 3 + consideration of LITS

In Figure D.11, the model response of the simplified linear-elastic material behavior (Model 1) is compared to experimental results presented in [68], showing the vertical and horizontal deformation at selected positions on the frame (middle of the frame ceiling and the frame corner). For this purpose, the measured deformation taken from [68], is corrected by both settlement and time-dependent behavior (creep of soil and concrete frame) prior to the fire experiment. Furthermore, the rotation of Corner 2 (connection of side wall and ceiling) was considered in the experimentally-obtained deformation (see Figure D.10). Hence, the displacements were corrected by additional displacements $(\Delta Y, \Delta Z)$, given by

$$\Delta Y = -d \sin(\varphi_2) \quad \text{and} \quad \Delta Z = d\,(1 - \cos(\varphi_2)), \qquad (D.3)$$

where d [m] is the distance of the mirror from the centroidal axis of the cross-section (approximately 0.39 m) and φ_2 [°] is the monitored rotation of Corner 2 (see [68] for details).

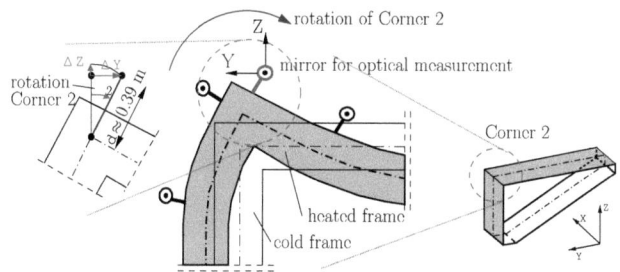

Figure D.10: Influence of rotation of frame corner on the deformations obtained in the experiments (see [68] for details)

The results obtained from Model 1 show only marginal changes in deformations during fire loading significantly underestimating the experimentally-observed deformations. Similar to the deformations, the rotations of the frame corners (see Figure D.12(a)) are underestimated. A stress distribution over the

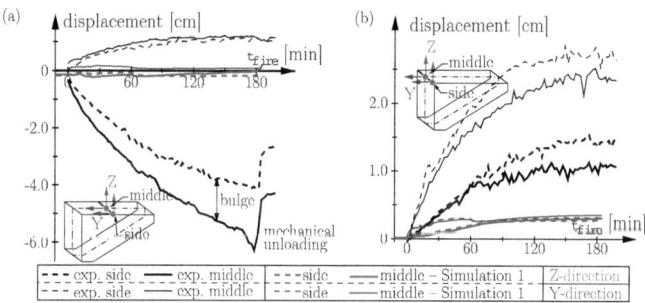

Figure D.11: Numerical results obtained from Model 1 for Frame 1: Deformations at (a) mid-span and (b) frame corner

cross-section is presented in Figure D.12(b) for a single finite element at mid-span. According to the underlying linear-elastic material behavior, the stresses show a linear distribution over the cross-section illustrating the fact that in this model the concrete stresses are not limited by the compressive/tensile strength. The peaks in the stress distribution refer to the reinforcement layers.

The distribution of internal forces over the frame cross-section is presented in Figure D.13 for different time instants (t_{fire} = 0, 30, 60, 120, and 180 min),

107

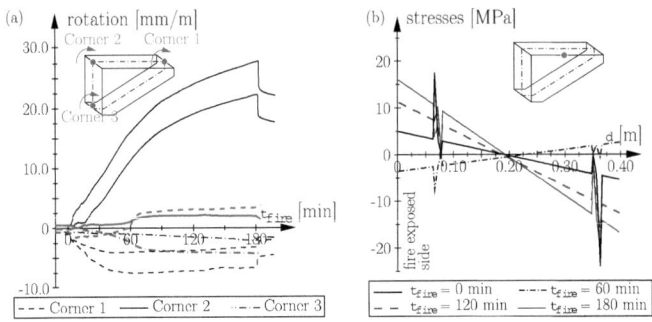

Figure D.12: Numerical results obtained from Model 1 for Frame 1: (a) Rotation of frame corners and (b) stress evolution at mid-span at different time instants ($t_{fire} = 0$, 60, 120, and 180 min)

obtained by averaging over the six finite elements along the longitudinal direction. In the ceiling, an increase in compression is observed with progressing fire loading, whereas the bottom of the frame is subjected to tensile loading. Up to 30 min of fire loading, the bending moments are increasing, followed by a decrease until the end of the simulation. This behavior is explained by the evolution of the equivalent temperature distribution given in Figure D.7.

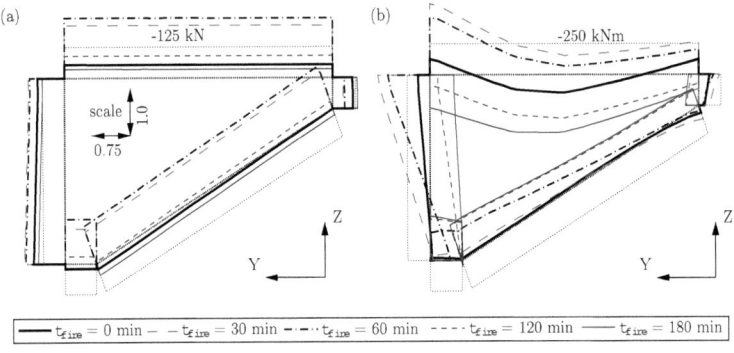

Figure D.13: Numerical results obtained from Model 1 for Frame 1: Distribution of (a) normal force and (b) bending moment for different time instants ($t_{fire} = 0$, 30, 60, 120, and 180 min)

Figure D.14 shows the horizontal and vertical deformation of the ceiling and the frame corner obtained from Model 2, giving deformations exceeding the experimental results. A bulge of up to 20 mm is observed, which is in good

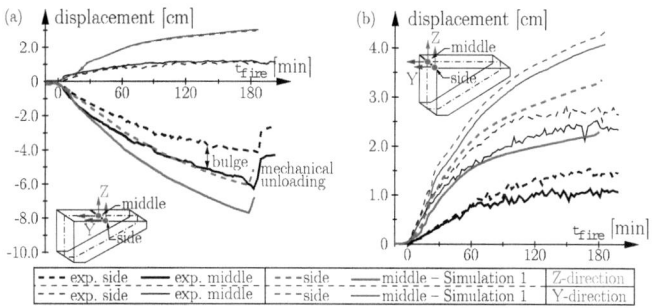

Figure D.14: Numerical results obtained from Model 2 for Frame 1: Deformations at (a) mid-span and (b) frame corner

agreement with the experiments (see Figure D.14(a)). The rotations at Corner 2, which are more than twice the rotations observed in the experiment (see Figure D.15(a)), result from the displacements in the ceiling of the frame in combination with plastic regions (see Figure D.16) developing at the side wall below the corner. As regards Corners 1 and 3, the obtained rotations are close to the experimental results.

The stress distributions presented in Figure D.15(b) are highly non-linear caused by the realistic consideration of both temperature loading and non-linear material behavior in Model 2. The limit of tensile loading of concrete

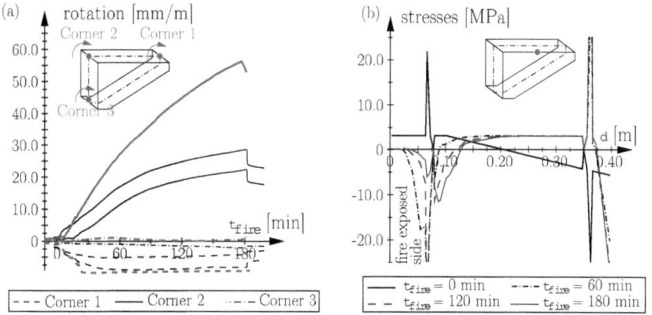

Figure D.15: Numerical results obtained from Model 2 for Frame 1: (a) Rotation of frame corners and (b) stress evolution at mid-span at different time instants ($t_{fire} = 0$, 60, 120, and 180 min)

is clearly identified at $f_t = 3$ MPa). Furthermore, the reduction of compressive stresses at the fire-exposed side is explained by material degradation due to high temperature loading in this region. Compared to the results obtained from

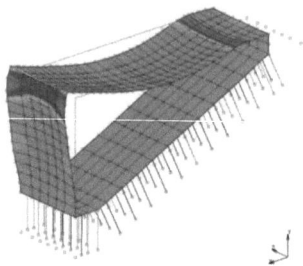

Figure D.16: Numerical results obtained from Model 2 for Frame 1: Plastic zones in outer reinforcement layer in circumferential direction (MSC.Marc Mentat)

Model 1, the increase in normal force is less pronounced (see Figure D.17(a)), which is explained by the underlying elasto-plastic material behavior. As regards the distribution of bending moments (see Figure D.17(b)), the results obtained from Model 2 show a redistribution of moments with progressing fire loading towards the parts of the frame with lower level of loading (Corners 1 and 3).

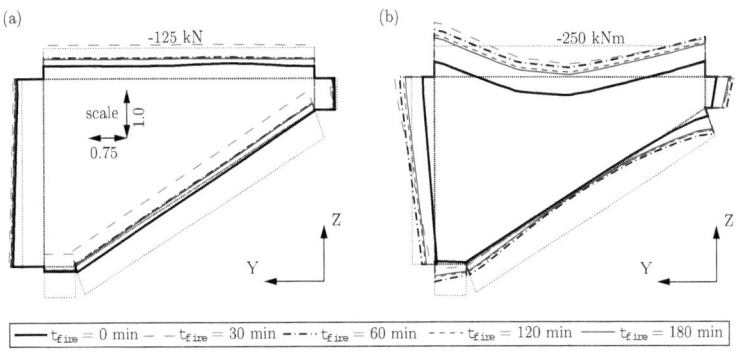

Figure D.17: Numerical results obtained from Model 2 for Frame 1: Distribution of (a) normal force and (b) bending moment for different time instants (t_{fire} = 0, 30, 60, 120, and 180 min)

In the analysis based on Model 3, the measured final spalling depth (see Figure D.18 and [68] for details) is considered by shifting the temperature distribution in the analysis to the final spalling depth. Consideration of spalling

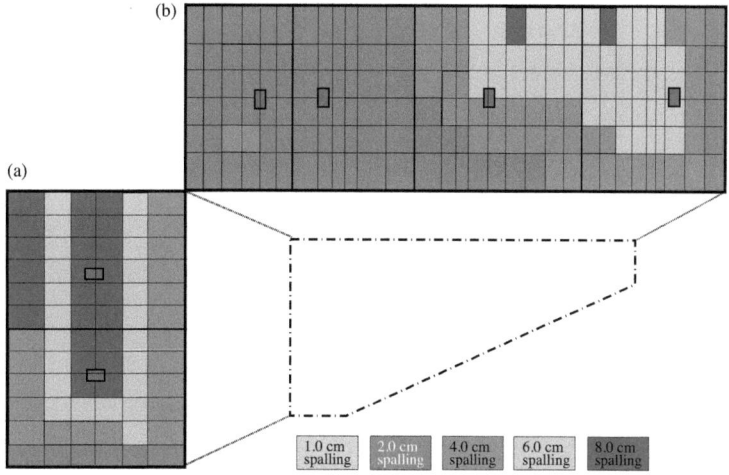

Figure D.18: Measured final spalling depth for (a) side wall and (b) ceiling of Frame 1 (see [68] for details)

resulted in vertical deformations presented in Figure D.19(a), showing good agreement with the experimental results. The horizontal deformations, however, still exceed the experimental observations. The deformations in vertical and horizontal direction in the frame corner (see Figure D.19(b)) are slightly reduced in comparison to Model 2 (see Figure D.14(b)). Due to the reduced

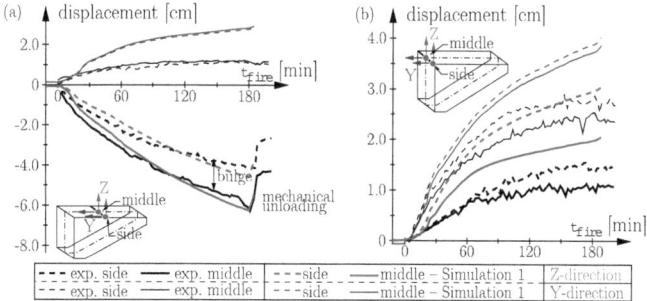

Figure D.19: Numerical results obtained from Model 3 for Frame 1: Deformations at (a) mid-span and (b) frame corner

thermal restraint, the rotations in Corner 2 are closer to the experimental re-

111

sults but still exceed the experiments which becomes more pronounced towards the end of the experiment (similarly to Model 2). The distribution of internal stress resultants is similar to the presented distributions for Model 2, which may be explained by the almost uniform nature of spalling. Localized spalling, on the other hand, might certainly affect the distribution of stress resultants. Accordingly, a slight increase of bending moment is observed in Corner 2 due to the reduced amount of spalling in this corner.

Finally, load induced thermal strains (LITS) are considered in Model 4. LITS is associated with combined thermo-mechanical loading of concrete as demonstrated in various experimental studies (see, e.g. [8, 70, 73]).

Recently, a differential strain formulation using a modified approach based on the model for $lits$ presented in [75] was proposed in order to describe the experimentally-observed behavior [67]. Starting from the strain composition of

$$\Delta = \Delta^{el}(T,) + \Delta^{th}(T) + \Delta^{lits}(T,), \tag{D.4}$$

where Δ^{el} and Δ^{th} represent the elastic strain and the free thermal strain, respectively, one can get determine Δ^{lits} in the strain rate formulation as

$$\Delta^{lits} = \left(\mathbb{D}_{n+1}^{lits} - \mathbb{D}_{n}^{lits} \right) : {}_{n+1}, \tag{D.5}$$

where $\mathbb{D}^{lits} = k\mathbb{I}^{vol}\ {}^{th}(T)/f_c(T)$ is the compliance tensor for the LITS component and ${}_{n+1}$ [MPa] referring to the actual load level of concrete in increment $n + 1$. With k [–], a factor introduced in [75], set to 0.4, and $f_c(T)$ [MPa] as the temperature-dependent compressive strength (see [67] for details).

Figure D.20 shows the influence of mechanical loading on the evolution of axial strain for an temperature increase up to 800 °C (see [67] for details). A decrease of thermal expansion with increasing mechanical load is observed.

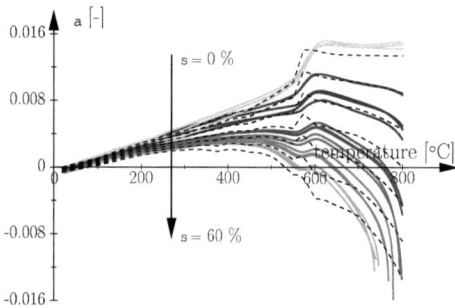

Figure D.20: Axial strain of concrete for different load levels $s = \sigma/f_{c,0} \cdot 100$ [–] (0-60 %) compared to experimental results (with $\Delta_a = \Delta_a^{el} + \Delta_a^{th} + \Delta_a^{lits}$, see [67] for details)

The influence of LITS on the numerically-obtained evolution of deformations is shown in Figure D.21. Compared to the previous results (see Figure D.19), a significant improvement in the quality of the simulation results is observed. Especially the vertical and horizontal deformations in the frame corner (see Figure D.21(b)) show good agreement with the experiment. The deformation in the mid-span of the ceiling is slightly exceeding the experimental results. The effect

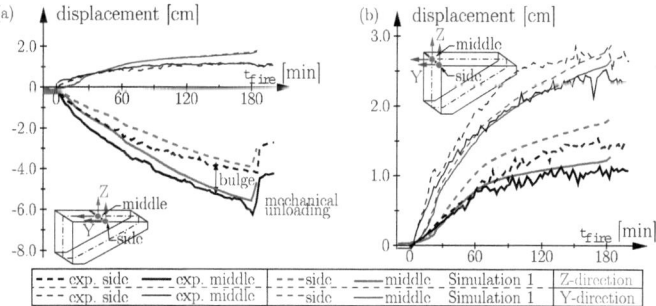

Figure D.21: Numerical results obtained from Model 4 for Frame 1: Deformations at (a) mid-span and (b) frame corner

of LITS is highlighted in the stress distribution presented in Figure D.22(b), showing reduced compressive stresses in comparison to Figure D.15(b). Furthermore, the stress reduction associated with LITS leads to lower internal forces (bending moment and normal force are reduced in comparison to Model 2, see

Figure D.17).

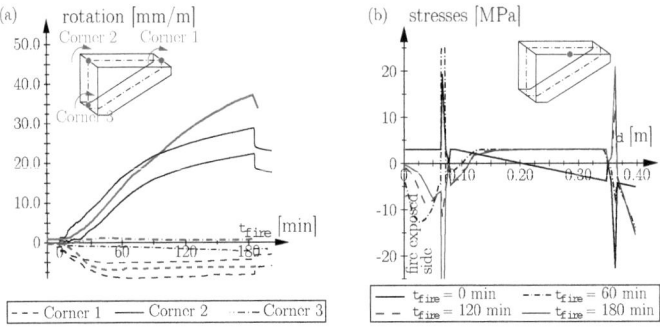

Figure D.22: Numerical results obtained from Model 4 for Frame 1: (a) Rotation of frame corners and (b) stress evolution at mid-span at different time instants (t_{fire} = 0, 60, 120, and 180 min)

D.5 Concluding remarks

In this paper, the numerical response of a reinforced-concrete frame under fire loading is compared to experimentally-observed data presented in [68]. In total, four different models with increasing complexity were considered:

Starting point for the simulation was the so-called equivalent temperature concept leading to a severe underestimation of deformations and, thus, rotations of frame corners. Consideration of elasto-plastic material response for concrete and steel led to a significant increase of deformations and rotations, resulting in an overestimation of the structural compliance. Finally, consideration of (i) spalling and (ii) the thermo-mechanical effect on the evolution of thermal strains led to a continuous increase in quality of the obtained numerical results and the agreement with experimental data (see Figures D.23 and D.24).

In the course of an additional simulation, the tensile strength was set to zero according to [20]. This change in the model had a severe influence on the model response, overestimating rotations and deformations when disregarding

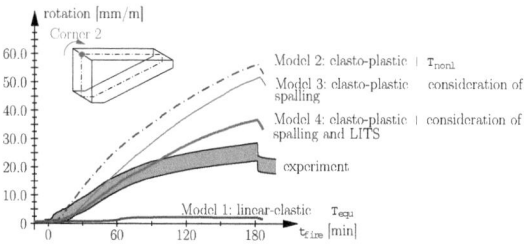

Figure D.23: Rotation at Corner 2 of Frame 1: Comparison of numerical results obtained from different models with large-scale experiment

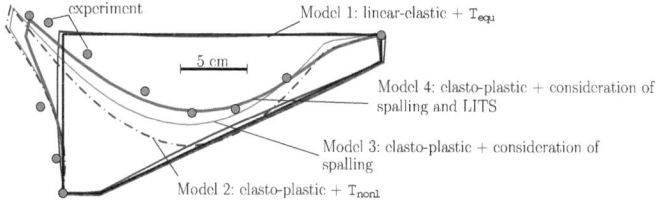

Figure D.24: Comparison of deformation of the whole frame in the symmetry plane for different material models with the experimental results at $t_{fire} = 180$ min for Frame 1

the tensile strength of concrete.

Further refinement of the temperature field in the simulations, especially in the frame corners, e.g. by means of thermal simulations, as well as consideration of softening behavior for concrete under tensile loading may lead to even better agreement between numerical results and experimental data.

The comparison between experimental and numerical results highlights, the conclusion that only in case load-induced thermal strains (LITS) and spalling are taken into account in the numerical simulation, realistic results for the behavior of concrete structures under fire loading can be accomplished.

Publication E

Der Einfluss des Rechen- und Materialmodells auf die Strukturantwort bei der Simulation von Tunnel unter Brandbelastung [65]

Authored by Thomas Ring, Christoph Wikete, Hannes Kari, Matthias Zeiml and Roman Lackner

To be published in: "Der Bauingenieur", 2012

Tunnelbrände sowie Brandversuche an Tragstrukturen haben aufgezeigt, dass Betontragwerke extremen physikalischen Prozessen (z.B. Abplatzungen, Abnahme des Elastizitätsmodul bzw. der Druckfestigkeit, usw. [66]) ausgesetzt sind, welche das Gesamttragverhalten wesentlich beeinflussen. In dieser Arbeit wird das Strukturverhalten von verschiedenen Tunnelgeometrien (Rechtecks- und Gewölbequerschnitt der offenen Bauweise) mithilfe eines FE-basierten Simulationstools untersucht. Bei diesen Studien werden verschiedene Berech-nungs- und Materialmodelle miteinander verglichen, begonnen mit einer linear-elastischen Simulation in Kombination mit einer linearen Temperaturverteilung über die Schalendicke (sogenannte äquivalente Temperatur). In einem Zwischenschritt wird elasto-plastisches Materialverhalten gemeinsam mit der linearen Temperaturverteilung (äquivalente Temperatur) angesetzt. Abschließend wird nichtlineares (elasto-plastisches) Materialverhalten und die reale, nichtlineare Temperaturverteilung berücksichtigt. Neben dem Einfluss der Berechnungsmethode wird der Einfluss der temperaturabhängigen Materialparameter (E, f_c) auf die Entwicklung des thermischen Zwangs untersucht. Hierbei werden die Designkurven für E und f_c gemäß Eurocode 2 [20] bzw. CEB [11] angesetzt. Der Vergleich der Ergebnisse zeigt einen großen Einfluss der angesetzten, normgemäßen Materialparameter [11, 20] auf die Simulation. Des Weiteren wurden große Unterschiede zwischen den Rechenmodellen hinsichtlich der Schnittgrößen und Verformungen beobachtet. Im Falle eines stark vereinfachten Rechenmodells (d.h. linear-elastisches Materialverhalten + lineare Temperaturverteilung) wurden die Schnittgrößen zufolge Zwang stark überschätzt und die Verformungen unterschätzt. Für die Prognose der Verformungen der Betonstruktur ist es unerlässlich ein realistisches Modell (d.h. elasto-plastisches Materialverhalten + nichtlineare Temperaturverteilung) zu wählen. Die realistischere Bestimmung der Schnittgrößen führt in der Praxis zu einer wirtschaftlich optimierten Bemessung von Tragstrukturen im Brandfall. Der Vergleich zwischen den einzelnen Tunnelgeometrien zeigt die Stärken und Schwächen der einzelnen Querschnitte (Rechtecks- bzw. Gewölbequerschnitt) unter Brandbeanspruchung auf.

Recent tunnel fires as well as fire experiments have revealed various physical processes (such as spalling, material degradation) which significantly reduce the load-bearing capacity of reinforced concrete structures under fire. Within this paper, the structural behavior of different tunnel geometries (rectangular and arched cross-section) is analyzed using a numerical FE-based tool. Different numerical and material models are compared, beginning with a linear-elastic material model together with a linear temperature distribution across the lining thickness (so-called equivalent temperature). In an intermediate step, elasto-plastic material behavior together with the linear temperature distribution (equivalent temperature) is applied. Finally, these simplified models are compared to a more sophisticated model, assuming elasto-plastic material behavior together with the real nonlinear temperature distribution. In addition to the influence of the numerical model, the influence of degradation of material parameters (E, f_c) using different standards (Eurocode 2 [20], CEB [11]) on the magnitude of the thermal restraint is investigated, showing a large influence of the used design curves for E and f_c. Moreover, a large influence of the underlying model on internal forces as well as deformations was observed. Considering a simplified numerical model (linear-elastic material behavior + linear temperature distribution) leads to an overestimation of the thermal restraint and an underestimation of the deformations. In order to obtain realistic deformations of concrete structures subjected to fire loading, the use of a realistic model (elasto-plastic material behavior + nonlinear temperature distribution) is required. The more realistic determination of internal stress resultants leads to a more economical design of carrying structures subjected to fire loading. The comparison of different tunnel geometries highlights the strengths and weaknesses of the investigated cross-sections (rectangular, arched) under fire loading.

E.1 Einleitung

In der Literatur sind einige Arbeiten zu finden, die sich mit dem Strukturverhalten von Tunnelschalen im Brandfall befassen. In [21] werden 3D Kontinuum-Elemente verwendet, wodurch eine große Anzahl an finiten Elementen benötigt wird, um die Tunnelschale zu diskretisieren. In [17] wird das Strukturverhalten (Spannungen, Verformungen) einer Innenschale aus Stahl in einem Tunnel unter Feuerlast behandelt. Traglastuntersuchungen unter Brandeinwirkung werden in [78] präsentiert, wobei Kreis- und Rechtecksquerschnitte untersucht wurden. Der Einfluss der Wahl des Materialmodells (linear-elastisch oder elasto-plastisch) auf das Strukturverhalten im Brandfall wurde bislang nicht detailliert behandelt und ist der Hauptgegenstand dieser Arbeit. Die Bemessung von Betontragwerken für den Brandfall wird in der Ingenieurpraxis zumeist mit einem vereinfachten Berechnungsmodell durchgeführt, welches neben linear-elastischem Materialverhalten auch eine lineare Temperaturverteilung über den Querschnitt ansetzt. Letztere wird als sogenannte äquivalente Temperatur bezeichnet und mittels Gleichgewicht der Schnittgrößen an einem eingespannten Balken ermittelt (siehe Abbildung E.1). Sie setzt sich aus zwei Anteilen zusammen, einer über den Querschnitt konstanten Temperaturerhöhung T_m [°C] und einem über den Querschnitt linearen Temperaturgradienten ΔT [°C/m] (für Details siehe [39]). Zusätzlich wird ein äquivalenter E-Modul E_{equ} [MPa] ermittelt. Durch diese Vereinfachung ist es möglich die Bemessung mithilfe von weit verbreiteten Stabwerksprogrammen durchzuführen.

Aufbauend auf die Erkenntnisse aus [71] wurden für die durchgeführten FE-Simulationen dicke Lagenelemente verwendet (siehe Abbildung E.2). Die Bewehrung wurde in einzelne Stahllagen mit einer Dicke äquivalent zur Quer-

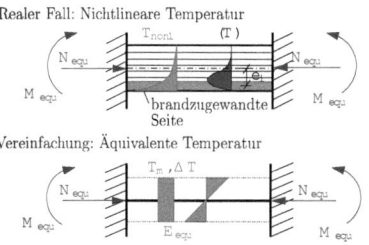

Figure E.1: Ermittlung der äquivalenten Temperatur [39]

Schnittsfläche der Bewehrung umgerechnet. Das Lagenelement hat die Vorteile, dass jeder einzelnen Lage ein Material (Beton oder Stahl) sowie eine Temperatur zugewiesen werden kann. Dadurch kann eine nichtlineare Temperaturverteilung als Belastung aufgebracht werden und diese bei der Zuweisung der (temperaturabhängigen) Materialparameter berücksichtigt werden. Weitere Vorteile sind die Möglichkeit zur Berücksichtigung von Abplatzungen durch Deaktivierung einzelner Lagen sowie die im Vergleich zum 3D-Kontinuum-Modell reduzierte Anzahl an benötigten finiten Elementen zur Berücksichtigung der nichtlinearen Temperaturverteilung (reduzierte Simulationsdauer).

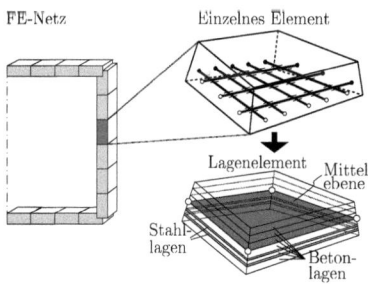

Figure E.2: Verwendetes numerisches Modell (vom FE-Netz zum Lagenelement [71])

Diese Arbeit gliedert sich wie folgt: Im Abschnitt E.2 werden Simulationsparameter (Geometrie, Belastung, Materialmodelle, Materialparameter) genauer definiert. Abschnitt E.3 zeigt ausgewählte Ergebnisse für die untersuchten Tunnelquerschnitte (Rechteck- und Gewölbequerschnitt) sowie einen Vergleich von Deformationen und Schnittgrößen für folgende Modelle:

1. lineare Simulation: linear-elastisches Materialverhalten + äquivalente Temperatur

2. Zwischenschritt: elasto-plastisches Materialverhalten + äquivalente Temperatur

3. nichtlineare Simulation: elasto-plastisches Materialverhalten + nichtlineare Temperatur

Des Weiteren wird in Abschnitt E.3 der Einfluss der temperaturabhängigen Materialparameter nach Eurocode 2 [20] bzw. CEB [11] dargestellt. Die Abschnitte E.4 und E.5 enthalten Schlussfolgerungen auf Basis der präsentierten Ergebnisse, und einen Ausblick auf geplante Aktivitäten.

E.2   Simulationsparameter

Für die durchgeführten Simulationen wurde Beton der Güte C25/30 (56)/BS1A[1] sowie Bewehrungsstahl der Güte BSt 550/M berücksichtigt. Es wurde jeweils ein 1 m breiter Tunnelstreifen simuliert, welcher an der Bodenplatte in vertikaler Richtung und im rechten unteren Eckpunkt in horizontaler Richtung gelagert ist. In Tunnellängsrichtung ist der Querschnitt ebenfalls unverschieblich gelagert (behinderte thermische Ausdehnung).

E.2.1   Geometrie und Belastung des Rechtecksquerschnitts

In Abbildung E.3 ist die rechte Hälfte des symmetrischen Rechtecksquerschnitts dargestellt. Neben den Abmessungen sind auch Angaben zum Bewehrungs-

---
[1] Gemäß ÖBV-Richtlinie "Wasserundurchlässige Betonbauwerke - Weiße Wannen" [52].

gehalt, zur Betondeckung und zu den Randbedingungen angeführt. Die Bewehrung in Umfangsrichtung ist in sieben Abschnitte geteilt. Die Dicke des Betonrahmens liegt zwischen 64 cm (Feldbereich) und 90 cm (Vouten im Eckbereich). Die Längsbewehrung beträgt über den gesamten Querschnitt 10.4 cm^2/m. Die mechanische Belastung des Rechtecksquerschnitts ist in Abbildung E.4 dargestellt. Sie setzt sich aus Anteilen aus Erdlast bzw. Erddruck, Verkehrslast und Wasserdruck zusammen. In der Simulation wird ein Lastfall untersucht, der durch eine asymmetrische Erdlast charakterisiert ist.

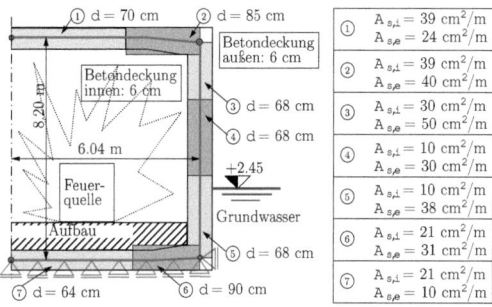

Figure E.3: **Rechtecksquerschnitt: Geometrie, Bewehrungsgehalt, Randbedingungen**

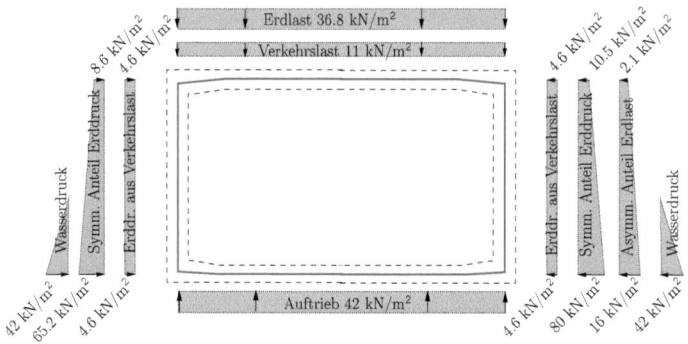

Figure E.4: **Rechtecksquerschnitt: mechanische Belastung**

E.2.2 Geometrie und Belastung des Gewölbequerschnitts

In Abbildung E.5 ist die rechte Hälfte des symmetrischen Gewölbequerschnitts dargestellt. Neben den Abmessungen sind auch Angaben zum Bewehrungsgehalt, zur Betondeckung und zu den Randbedingungen angeführt. Die Bewehrung in Umfangsrichtung ist in fünf Abschnitte geteilt. Die Dicke der Betonschale liegt zwischen 60 cm (im Gewölbe) und 70 cm (Ecken in der Bodenplatte). Die Längsbewehrung beträgt über den gesamten Querschnitt 7.5 cm^2/m. Die mechanische Belastung des Gewölbequerschnitts ist in Abbildung E.6 dargestellt. Sie setzt sich aus Anteilen aus Erdlast bzw. Erddruck[2], Verkehrslast und Wasserdruck zusammen. In der Simulation wird ein Lastfall untersucht, der durch eine asymmetrische Erdlast charakterisiert ist.

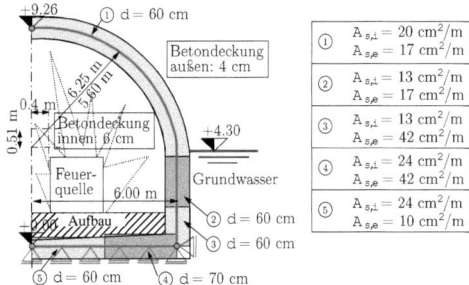

Figure E.5: Gewölbequerschnitt: Geometrie, Bewehrungsgehalt, Randbedingungen

E.2.3 Thermische Belastung

Die Temperatureindringkurven zu unterschiedlichen Zeitpunkten wurden durch Simulation des Wärmetransports in den Beton [81] unter Verwendung der thermischen Parameter gemäß Eurocode 1 und 2 [18, 20] ermittelt. Hierbei wurde eine für Tunnelbrände repräsentative Temperatur-Zeitkurve angesetzt (siehe

---

[2] Die Erdlast wurde vereinfachend konstant gehalten, womit eine Veränderung des Erddrucks aufgrund der Ausdehnung des Tragwerks nicht berücksichtigt wurde.

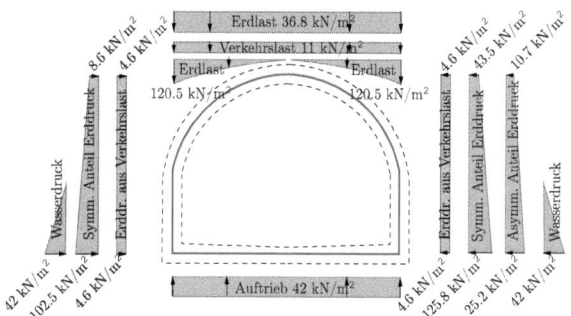

Figure E.6: Gewölbequerschnitt: mechanische Belastung

Abbildung E.7). Die Temperatureinwirkung wurde vereinfachend gleichmäßig auf jenen Teil des Tunnelquerschnitts angesetzt, der nicht durch den Aufbau im Tunnel (z.B. Gleisbett, Straßenaufbau) geschützt wird[3]. Dementsprechend wurden die Bodenplatte und der untere Teil der Wände (bis 1 m über FUK) thermisch nicht belastet.

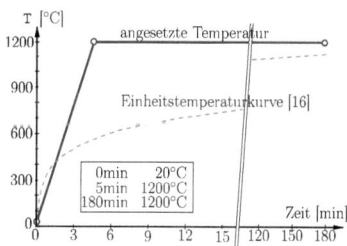

Figure E.7: Angesetzte Temperatur-Zeitkurve

E.2.4 Materialmodelle

In diesem Beitrag wurden die folgenden Rechen- bzw. Materialmodelle untersucht:

---
[3]Für weitere Informationen zur Ermittlung einer realistischeren Verteilung der Brandraumtemperaturen mithilfe von CFD-Simulationen siehe [5, 6].

- **Lineare Simulation:**

  In der linearen Simulation wurde neben einem linear-elastischen Materialverhalten (siehe Abbildung E.8) die äquivalente Temperaturverteilung (siehe Abschnitt E.2.3) angesetzt. Die Abnahme des E-Moduls und der Festigkeit wird bei der Ermittlung der äquivalenten Temperatur berücksichtigt (siehe Abschnitt E.2.5). In der Struktursimulation wird ein äquivalenter E-Modul gleichmäßig über die Querschnittsdicke angesetzt, die Spannungen in Beton und Bewehrung werden nicht begrenzt, weder im Druck- noch im Zugbereich. Für Beton und Stahl wurden konstante Wärmeausdehnungskoeffizienten verwendet ($_B = 1.0 \cdot 10^{-5}$ [K^{-1}], $_S = 1.2 \cdot 10^{-5}$ [K^{-1}]).

- **Zwischenschritt:**

  Hierbei wird nichtlineares (elasto-plastisches) Materialverhalten (siehe Abbildung E.8) zusammen mit der äquivalenten Temperatur angesetzt. Die

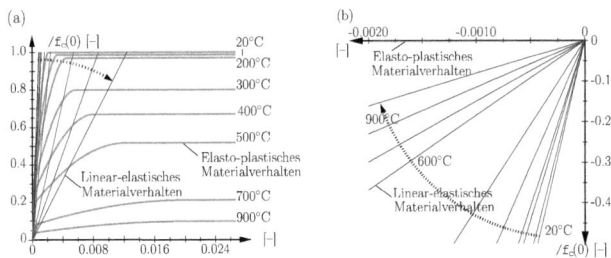

Figure E.8: Materialverhalten von Beton: Spannungs-Dehnungsbeziehungen (a) im Druckbereich und (b) im Zugbereich [69]

Zugfestigkeit f_t wird gemäß Eurocode 2-1-2 [20] vernachlässigt (siehe Abbildung E.8(b)). Die Druckfestigkeit bei Raumtemperatur, $f_c(0)$, wurde mit 25 MPa angenommen. Gemäß Eurocode 2-1-2 [20] wurde die Wärmeausdehnung von Beton mit quarzitischen Zuschlägen als nichtlineare Funktion der Temperatur angesetzt. Für den Bewehrungsstahl wurde ebenfalls elasto-plastisches Materialverhalten mit einer Fließspannung bei Raumtemperatur von $f_y(0)$ = 550 MPa und einer temperaturabhängigen Abnahme gemäß Eurocode 2-1-2 [20] angesetzt.

- Nichtlineare Simulation:
  Hierbei wurde das bereits im Zwischenschritt verwendete elasto-plastische Materialmodell gemäß Eurocode 2-1-2 [20] gemeinsam mit einer nichtlinearen Temperaturverteilung (Berechnung siehe Abschnitt E.2.3) angesetzt. Durch Verwendung der Schalenelemente (siehe Abschnitt E.1) kann die nichtlineare Temperaturverteilung in der Simulation direkt berücksichtigt werden, wodurch eine möglichst realitätsnahe Simulation des Verhaltens von Betonstrukturen unter Feuerlast ermöglicht wird.

E.2.5 Temperaturabhängigkeit der Materialparameter

Um den Einfluss der Temperaturabhängigkeit der Materialparameter auf den thermischen Zwang bzw. die Simulationsergebnisse zu untersuchen, werden unterschiedliche Designkurven für die Materialparameter von Beton (E, f_c) angesetzt (siehe Abbildung E.9). Der Abfall der Druckfestigkeiten gemäß Eurocode 2-1-2 [20] und CEB [11] ist ähnlich, ab einer Temperatur von 350 °C ergibt sich gemäß CEB ein schnellerer Abfall der Druckfestigkeit. Die Entwicklung des E-Moduls zeigt allerdings größere Unterschiede. Im Falle des Eurocode 2-1-2 [20] wurde die Entwicklung durch Rückrechnung der dort gegebenen Spannungs-Dehnungsbeziehungen ermittelt, mit einem E-Modul bei Raumtemperatur von $E(0)$ = 15 GPa. Dies unterscheidet sich wesentlich vom E-Modul für C25/30 gemäß Eurocode 2-1-2 [19] von $E(0)$ = 31 GPa. Dieser Wert wurde für $E(0)$ bei einer temperaturabhängigen Abnahme des E-Moduls gemäß CEB [11], die darüber hinaus wesentlich langsamer ausfällt, angesetzt. Aus diesen unterschiedlichen Ansätzen ergeben sich wesentliche Unterschiede bei der äquivalenten Temperatur (siehe Abschnitt E.3).

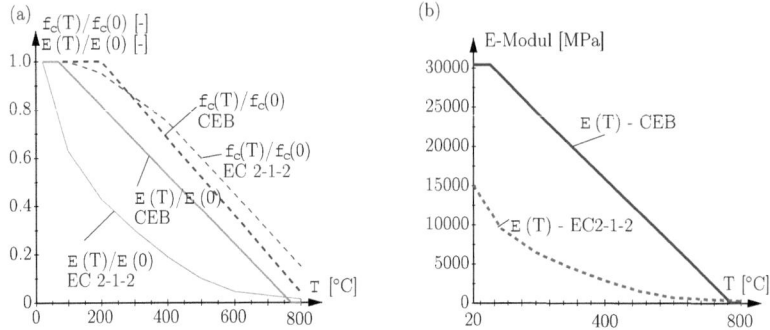

Figure E.9: (a) Temperaturabhängige Materialparameter [11, 20]; (b) Temperaturabhängige Abnahme des E-Moduls (Ansatz nach EC2-1-2 [20] und CEB [11])

E.3 Ergebnisse

Durch Gegenüberstellung ausgewählter numerischer Ergebnisse (Verläufe von Spannungen, Biegemoment, Normalkraft sowie Verformungen) wird die Auswirkung der unterschiedlichen Rechen- und Materialmodelle auf die Simulationsergebnisse bewertet. Außerdem werden die Ergebnisse an ausgewählten Punkten (siehe Abbildung E.10) in Tabellenform angegeben. Die ausgewählten Punkte stellen Extremwerte von Schnittgrößen bzw. Verformungen dar. Beim Rechtecksquerschnitt (siehe Abbildung E.10 (a)) wurden hierfür die rechte Rahmenecke im Anschlussbereich Bodenplatte/Wand, die Feldmitte der Tunneldecke und das linke Rahmeneck Tunneldecke/Wand gewählt. Aufgrund der asymmetrischen mechanischen Belastung ergeben sich in diesen Punkten beispielsweise die größten Biegemomente. Beim Gewölbequerschnitt wird neben der rechten unteren Rahmenecke und der Tunnelfirste auch ein Punkt in der linken Tunnelschulter für den Vergleich herangezogen (siehe Abbildung E.10 (b)).

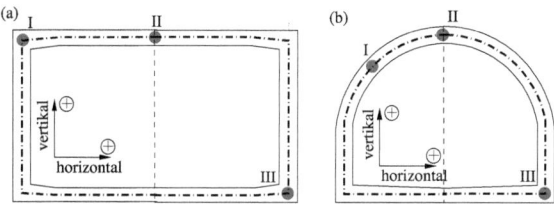

Figure E.10: Gewählte Punkte für den Vergleich von Spannungen, Schnittgrößen und Verformungen: (a) Rechtecksquerschnitt, (b) Gewölbequerschnitt

E.3.1 Rechtecksquerschnitt

Abbildung E.11 zeigt die Spannungen in der linken oberen Ecke (Punkt I) für die lineare und die nichtlineare Simulation zu unterschiedlichen Zeitpunkten (t_{fire} = 0, 20, 60, 180 min). Die lineare Verteilung der Spannungen über den Querschnitt in Abbildung E.11(a) ergibt sich aus dem Ansatz von linearelastischem Materialverhalten zusammen mit der äquivalenten Temperatur (lineare Temperaturverteilung). Die Spannungsspitzen repräsentieren die jeweiligen Bewehrungsspannungen. Infolge der thermischen Beanspruchung nehmen

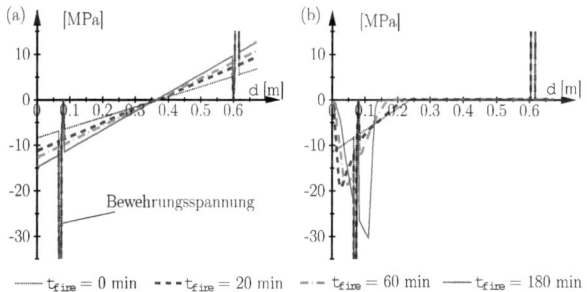

Figure E.11: Spannungen im Rechtecksquerschnitt in der linken oberen Ecke (Punkt I) für (a) die lineare Simulation und (b) die nichtlineare Simulation zu verschiedenen Zeitpunkten (t_{fire} = 0, 20, 60, 180 min)

die Spannungen zu. Es ist insbesondere zu beobachten, dass zu jedem Zeitpunkt die Zugspannungen die Zugfestigkeit des Betons überschreiten, eine bekannte

Schwäche der linear-elastischen Simulation. Im Falle der nichtlineren Simulation (siehe Abbildung E.11 (b)) treten die charakteristischen Druckspannungsglocken auf, die durch die nichtlineare Temperatureindringkurven und die temperaturabhängige Abnahme der Materialparameter begründet werden können. Mit Fortdauer der Brandbelastung verschiebt sich die Spannungsglocke zufolge der Temperatureindringung weiter in Richtung Querschnittsmitte. Des Weiteren erkennt man die Deaktivierung der Zugfestigkeit für Beton gemäß [20].

Der Verlauf des Biegemoments in Abbildung E.12 ist exemplarisch für die nichtlineare Simulation zu unterschiedlichen Zeitpunkten (t_{fire} = 0, 20, 60, 180 min) dargestellt. Die unverschiebliche Lagerung der Bodenplatte spiegelt sich

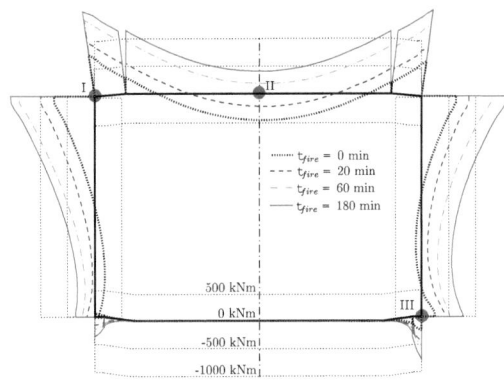

Figure E.12: Verlauf des Biegemoments über den Rechtecksquerschnitt für die nichtlineare Simulation zu verschiedenen Zeitpunkten (t_{fire} = 0, 20, 60, 180 min)

in den Ergebnissen wider. Mit zunehmender Branddauer steigt der thermische Zwang, dadurch nehmen die negativen Biegemomente über den gesamten Tunnelquerschnitt (insbesondere in den brandbelasteten Bereichen) zu, wohingegen die positiven Biegemomente abnehmen und es zu einem Umschlagen (z.B. in Deckenmitte) des Biegemoments kommt. In den beiden oberen Rahmenecken werden die höchsten Biegemomente beobachtet.

Abbildung E.13 zeigt exemplarisch den Verlauf der Normalkraft für die nichtlineare Simulation zu unterschiedlichen Zeitpunkten (t_{fire} = 0, 20, 60, 180 min). Es ist zu erkennen, dass der Brandfall auf die Normalkräfte nur einen geringen Einfluss hat. Im Bereich der Seitewände zeigt sich kaum eine Veränderung der Normalkraft, da sich die mechanische Belastung nicht ändert. Im Bereich der Tunneldecke kommt es infolge der Brand-einwirkung zu einer leichten Erhöhung und in der Tunnelsohle zu einer Reduktion der Normalkraft. Dies kann damit begründet werden, dass in der Decke thermischer Zwang entsteht, die Bodenplatte andererseits wird thermisch nicht belastet und dehnt sich daher nicht aus.

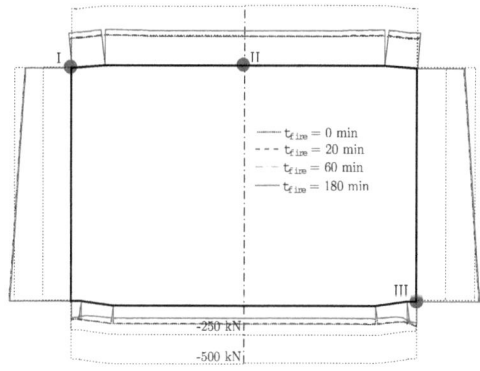

Figure E.13: Verlauf der Normalkraft über den Rechtecksquerschnitt für die nichtlineare Simulation zu verschiedenen Zeitpunkten (t_{fire} = 0, 20, 60, 180 min)

Abbildung E.14 zeigt einen Vergleich der Verformungen des Rechtecksquerschnitts für die lineare und die nichtlineare Simulation zu unterschiedlichen Zeitpunkten (t_{fire} = 0, 20, 60, 180 min). Bei der linearen Simulation ergeben sich sehr kleine Verformungen, die von der Branddauer annähernd unabhängig sind, wohingegen die nichtlineare Simulation Verformungen bis zu 6 cm in der Deckenmitte nach 180 min Branddauer prognostiziert. Ein Vergleich von exper-

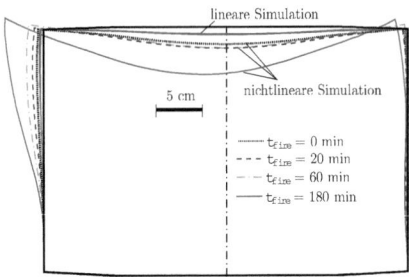

Figure E.14: Verformungen des Rechtecksquerschnitts: Vergleich der linearen mit der nichtlinearen Simulation zu verschiedenen Zeitpunkten (t_{fire} = 0, 20, 60, 180 min)

imentellen Ergebnissen mit numerisch ermittelten Verformungen zeigte, dass die mit einer linear-elastischen Simulation prognostizierten Verformungen zu gering sind um die Realität wiederzugeben [69]. Andererseits zeigten die Ergebnisse, die mit dem auch hier verwendeten, nichtlinearen Rechen- und Materialmodell ermittelt wurden, sehr gute Übereinstimmung mit den gemessenen Verformungen [69]. In den Verformungen der nichtlinearen Simulation sind die Auswirkungen der asymmetrischen Belastung sichtbar, der Tunnelquerschnitt wird auf der rechten Seitenwand höher belastet und verschiebt sich daher nach links. Die großen Horizontalverschiebungen in der nichtlinearen Simulation (Ausweichen des Tunnelquerschnitts) werden in der Realität durch Aktivierung des passiven Erddrucks begrenzt.

Ein tabellarischer Vergleich der Schnittgrößen (Biegemoment, Normalkraft) sowie der Verformungen zeigt den Einfluss der angesetzten Rechen- bzw. Materialmodelle auf die Ergebnisse (siehe Tabellen E.1 und E.2). Man erkennt, dass die Ergebnisse des Zwischenschritts (siehe Abschnitt E.2.4) und des nichtlinearen Materialmodells zum Zeitpunkt t_{fire} = 0 min ident sind, da sich diese Modelle nur in der Temperaturbelastung unterscheiden. Die Verformungen unter Ansatz der linearen Simulation sind vergleichsweise klein (ca. 50 % von jenen der nichtlinearen Simulation nach EC2-1-2 und ca. 30 % von jenen der nichtlinearen Simulation nach CEB). Die geringen Unterschiede zwischen den Simulationen mit Materialparametern nach EC2-1-2 [20] und CEB [11] sind auf den unterschiedlichen E-Modul bei Raumtemperatur zurückzuführen (EC2-1-2: $E(0) = 15$ GPa, CEB: $E(0) = 31$ GPa siehe Abschnitt E.2.5). Größere Un-

BIEGEMOMENTE		EC2-1-2			CEB	
	linear [kNm]	Zwischen-schritt [kNm]	nicht-linear [kNm]	linear [kNm]	Zwischen-schritt [kNm]	nicht-linear [kNm]
Punkt I	-749	-768	-768	-745	-768	-768
Punkt II	501	478	478	504	475	475
Punkt III	-267	-256	-256	-271	-254	-254
NORMALKRÄFTE		EC2-1-2			CEB	
	[kN]	[kN]	[kN]	[kN]	[kN]	[kN]
Punkt I (Decke)	-253	-257	-257	-252	-258	-258
Punkt I (Wand)	-417	-416	-416	-417	-416	-416
Punkt II	-241	-245	-245	-240	-246	-246
Punkt III (Bodenplatte)	-184	-180	-180	-185	-180	-180
Punkt III (Wand)	-535	-534	-534	-535	-535	-535
VERFORMUNGEN		EC2-1-2			CEB	
	[mm]	[mm]	[mm]	[mm]	[mm]	[mm]
Punkt I (vertikal)	-0.3	0.3	0.3	-0.2	0.6	0.6
Punkt I (horizontal)	-3.8	-7.8	-7.8	-1.7	-6.0	-6.0
Punkt II (vertikal)	-11.1	-22.9	-22.9	-6.2	-18.9	-18.9
Punkt II (horizontal)	-2.9	-5.8	-5.8	-1.6	-4.6	-4.6

Table E.1: **Vergleich der Ergebnisse an unterschiedlichen Punkten (siehe Abbildung E.10(a)) zum Zeitpunkt $t_{fire} = 0$ min für den Rechtecksquerschnitt**

terschiede in den Ergebnissen sind in den Schnittgrößen und Verformungen zum Zeitpunkt t_{fire} = 60 min zu erkennen (siehe Tabelle E.2). Im Vergleich zwischen linearer und nichtlinearer Simulation fällt auf, dass bei der nichtlinearen Simulation das Biegemoment in der rechten unteren Ecke (Punkt III) stark zunimmt. Diese Zunahme ist die Folge von plastischen Verformungen, die eine Umlagerung von den thermisch belasteten Bereichen hin zur thermisch nicht geschädigten unteren Rahmenecke ermöglichen. Diese Umlagerung spiegelt sich auch in den Verformungen wider. In der nichtlinearen Simulation werden vertikale Verformungen von 27.9 mm in Deckenmitte prognostiziert.

In der linearen Simulation ist andererseits kaum eine Veränderung der Verformungen zu erkennen (vgl. mit Tabelle E.1). Der Vergleich der Ergebnisse unter Ansatz der Materialparameter nach EC 2-1-2 [20] und CEB [11] zeigt wesentlich höhere Biegemomente, wenn die Parameter nach CEB angesetzt werden. Dies kann vor allem durch den höheren thermischen Zwang (siehe Abbildung E.15) erklärt werden, der durch einen höheren E-Modul (siehe Abbildung E.9(b)) verursacht wird. Dies hat wiederum eine höhere äquivalente Temperatur und somit größere Zwangsschnittgrößen zur Folge. Der Zwischenschritt prognos-

BIEGEMOMENTE	EC2-1-2			CEB		
	linear [kNm]	Zwischen-schritt [kNm]	nicht-linear [kNm]	linear [kNm]	Zwischen-schritt [kNm]	nicht-linear [kNm]
Punkt I	-1228	-1044	-1433	-1602	-1095	-1561
Punkt II	17	201	-186	-360	150	-314
Punkt III	-490	-414	-544	-694	-459	-744
NORMALKRÄFTE	EC2-1-2			CEB		
	[kN]	[kN]	[kN]	[kN]	[kN]	[kN]
Punkt I (Decke)	-285	-272	-303	-307	-272	-294
Punkt I (Wand)	-417	-417	-418	-417	-417	-417
Punkt II	-273	-260	-291	-295	-269	-283
Punkt III (Bodenplatte)	-171	-161	-172	-167	-195	-213
Punkt III (Wand)	-535	-533	-534	-535	-535	-534
VERFORMUNGEN	EC2-1-2			CEB		
	[mm]	[mm]	[mm]	[mm]	[mm]	[mm]
Punkt I (vertikal)	0.5	1.8	6.2	0.7	2.4	9.1
Punkt I (horizontal)	-4.7	-9.9	-19.9	-2.4	-9.5	-21.7
Punkt II (vertikal)	-9.9	-25.4	-27.9	-4.3	-22.1	-21.1
Punkt II (horizontal)	-3.3	-7.6	-11.5	-1.8	-7.2	-12.5

Table E.2: Vergleich der Ergebnisse an unterschiedlichen Punkten (siehe Abbildung E.10(a)) zum Zeitpunkt $t_{fire} = 60$ min für den Rechtecksquerschnitt

tiziert die betragsmäßig niedrigsten Biegemomente. Die Unterschätzung des thermischen Zwangs kann auf die doppelte Berücksichtigung von plastischem Materialverhalten zurückgeführt werden (einerseits wird die äquivalente Temperatur unter Ansatz nichtlinearer Materialparameter ermittelt, andererseits wird im Strukturmodell wiederum nichtlineares Materialverhalten angesetzt). Die Verformungen sind größer als jene der linearen Simulation, es sind jedoch

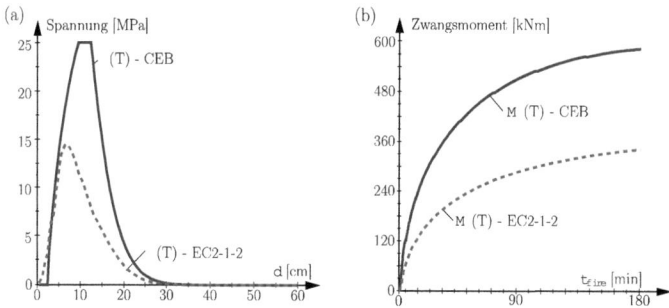

Figure E.15: Ermittlung der äquivalenten Temperatur nach EC2-1-2 [20] und CEB [11]: (a) Thermischer Zwang (Spannungsverteilung) zum Zeitpunkt t_{fire} = 60 min, (b) zeitliche Entwicklung des Zwangsmoments für eine Querschnittsdicke von 60 cm

immer noch große Abweichungen von den Ergebnissen der nichtlinearen Simulation zu beobachten.

E.3.2 Gewölbequerschnitt

Der Verlauf der Biegemomente in Abbildung E.16 ist exemplarisch für die nichtlineare Simulation zu unterschiedlichen Zeitpunkten (t_{fire} = 0, 20, 60, 180 min) dargestellt. Das größte Biegemoment im Gewölbequerschnitt bildet sich zufolge der asymmetrischen Erdlast an der rechten unteren Ecke (Punkt III) aus, die Werte in der Tunnelschulter (Punkt I) liegen knapp darunter. Aufgrund der Querschnittsform verteilen sind die Biegemomente jedoch unter Brandbelastung gleichmäßiger im Vergleich zum Rechtecks-querschnitt. Abbildung E.17 zeigt exemplarisch den Verlauf der Normalkraft für die nichtlineare Simulation zu unterschiedlichen Zeitpunkten (t_{fire} = 0, 20, 60, 180 min). Es ist wiederum zu erkennen, dass der Brandfall auf die Normalkräfte nur einen geringen Einfluss hat. Aufgrund der Gewölbetragwirkung des Querschnitts wird die Belastung der Tunnelschale über die Wände in den Untergrund abgeleitet. Die thermisch nicht belastete Bodenplatte ist nur geringfügig auf Druck belastet.

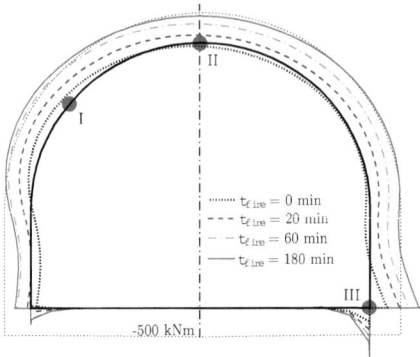

Figure E.16: Verlauf des Biegemoments über den Gewölbequerschnitt für die nichtlineare Simulation zu verschiedenen Zeitpunkten (t_{fire} = 0, 20, 60, 180 min)

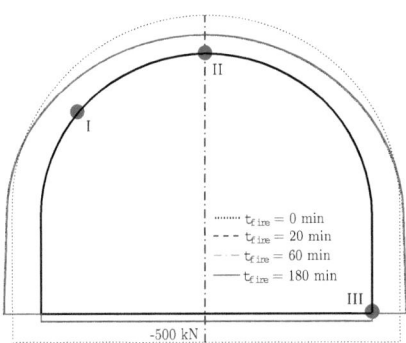

Figure E.17: Verlauf der Normalkraft über den Gewölbequerschnitt für die nichtlineare Simulation zu verschiedenen Zeitpunkten (t_{fire} = 0, 20, 60, 180 min)

Abbildung E.18 zeigt einen Vergleich der Verformungen des Gewölbequerschnitts für die lineare und die nichtlineare Simulation zu unterschiedlichen Zeitpunkten (t_{fire} = 0, 20, 60, 180 min). Bei der linearen Simulation ergeben sich wieder sehr kleine Verformungen. In der nichtlinearen Simulation für t_{fire} = 180 min ist hingegen ein starkes Anwachsen der Verformungen zu erkennen, das auf das Ausbilden eines plastischen Bereiches bei Punkt I zurückgeführt werden kann, in dem sowohl die innere als auch die äußere Bewehrungslage zu fließen beginnen. Es sind wiederum die Auswirkungen der asymmetrischen Belastung in den Verformungen der nichtlinearen Simulation sichtbar. Der Tunnelquerschnitt wird auf der rechten Seitenwand höher belastet und verschiebt sich daher nach links. Die vergleichsweise großen Horizontalverschiebungen in der nichtlinearen Simulation (Ausweichen des Gesamtquerschnitts) werden in der Realität durch Aktivierung des passiven Erddrucks begrenzt.

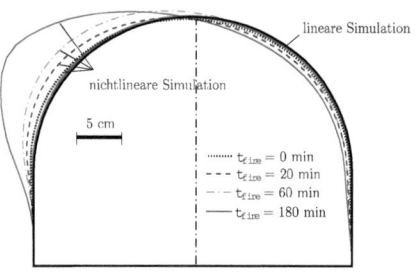

Figure E.18: Verformungen des Gewölbequerschnitts: Vergleich der linearen mit der nichtlinearen Simulation zu verschiedenen Zeitpunkten ($t_{fire} = 0, 20, 60, 180$ min)

Ein tabellarischer Vergleich der Schnittgrößen (Biegemoment, Normalkraft) sowie der Verformungen zeigt den Einfluss der angesetzten Rechen- bzw. Materialmodelle auf die Ergebnisse (siehe Tabellen E.3 bis E.4). Es können ähnliche Schlussfolgerungen wie beim Rechtecksquerschnitt gezogen werden. So ist nur ein geringer Einfluss des Rechenmodells (linear, nichtlinear) auf die Simulationsergebnisse vor Brandbelastung erkennbar. Wegen der Gewölbetragwirkung werden kleinere Biegemomente als im Rechtecksquerschnitt (außer in der unteren Rahmenecke, Punkt III) beobachtet. Die maximalen Verformungen (an Punkt I in horizontaler Richtung) betragen in der nichtlinearen Simulation nach EC2-1-1 8.6 mm und in der linearen Simulation 6.0 mm (siehe Tabelle E.3). Für die Simulationen nach CEB ergeben sich wieder aufgrund des höheren E-Moduls bei Raumtemperatur geringere Verformungen (etwa 60 % der Simulationsergebnisse nach EC2-1-2).

BIEGEMOMENTE	EC2-1-2			CEB		
	linear [kNm]	Zwischen-schritt [kNm]	nicht-linear [kNm]	linear [kNm]	Zwischen-schritt [kNm]	nicht-linear [kNm]
Punkt I	-93	-96	-96	-93	-97	-97
Punkt II	56	58	58	56	60	60
Punkt III	-350	-336	-336	-351	-324	-324
NORMALKRÄFTE	EC2-1-2			CEB		
	[kN]	[kN]	[kN]	[kN]	[kN]	[kN]
Punkt I	-427	-428	-428	-427	-429	-429
Punkt II	-331	-331	-331	-331	-332	-332
Punkt III (Bodenplatte)	-138	-138	-138	-139	-138	-138
Punkt III (Wand)	-614	-612	-612	-613	-611	-611
VERFORMUNGEN	EC2-1-2			CEB		
	[mm]	[mm]	[mm]	[mm]	[mm]	[mm]
Punkt I (vertikal)	0.4	0.5	0.5	0.2	0.3	0.3
Punkt I (horizontal)	-6.0	-8.6	-8.6	-3.3	-5.8	-5.8
Punkt II (vertikal)	-1.7	-3.5	-3.5	-0.9	-2.6	-2.6
Punkt II (horizontal)	-5.5	-7.2	-7.2	-3.0	-4.6	-4.6

Table E.3: Vergleich der Ergebnisse an unterschiedlichen Punkten (siehe Abbildung E.10(b)) zum Zeitpunkt $t_{fire} = 0$ min für den Gewölbequerschnitt

Größere Unterschiede in den Ergebnissen sind in den Schnittgrößen und Verformungen zum Zeitpunkt t_{fire} = 60 min zu erkennen (siehe Tabelle E.4). Der Vergleich der Biegemomente zeigt ein höheres Moment in der Tunnelschulter im Fall einer linearen Simulation, wobei in der unteren Rahmenecke (Punkt III) ein Extremwert im Biegemomentenverlauf in der nichtlinearen Simulation erzielt wird. Dies kann auf Umlagerungseffekte in der nichtlinearen Simulation von brandbelasteten in thermisch ungeschädigte Bereiche zurückgeführt werden. Wie bereits beim Rechtecksquerschnitt beobachtet, werden die Biegemomente im Zwischenschritt erneut unterschätzt. Die Normalkräfte ändern sich infolge der Brandbelastung kaum (vgl. mit Tabelle E.3). Nur bei der linearen Simulation kommt es zu einer geringfügigen Erhöhung der Normalkräfte im Schulter- und Firstbereich (Punkte I und II). Hinsichtlich der Verformungen sind große Unterschiede in den untersuchten Rechen- und Materialmodellen erkennbar. Die lineare Simulation prognostiziert sehr geringe Verformungen. Im Zwischenschritt werden Verformungen bis zum 3-fachen und in der nichtlinearen Simulation bis zum 10-fachen der linearen Simulation prognostiziert. (siehe z.B. die vertikalen Verformungen in Punkt I). Der Vergleich der Ergebnisse unter Ansatz der Materialparameter nach EC2-1-2 [20] und CEB [11] zeigt wiederum wesentlich höhere Biegemomente im letzteren Fall. Durch den höheren thermischen Zwang in den Simulationen mit Materialparametern nach CEB kommt es auch zu größeren Verformungen im Vergleich zu den Simulationen nach EC2-1-2.

BIEGEMOMENTE	EC2-1-2			CEB		
	linear [kNm]	Zwischen-schritt [kNm]	nicht-linear [kNm]	linear [kNm]	Zwischen-schritt [kNm]	nicht-linear [kNm]
Punkt I	-469	-260	-467	-726	-269	-560
Punkt II	-342	-129	-339	-612	-138	-426
Punkt III	-609	-547	-705	-762	-570	-816
NORMALKRÄFTE	EC2-1-2			CEB		
	[kN]	[kN]	[kN]	[kN]	[kN]	[kN]
Punkt I	-438	-425	-429	-448	-423	-426
Punkt II	-346	-332	-336	-358	-330	-333
Punkt III (Bodenplatte)	-123	-138	-134	-111	-139	-137
Punkt III (Wand)	-614	-618	-616	-614	-618	-617
VERFORMUNGEN	EC2-1-2			CEB		
	[mm]	[mm]	[mm]	[mm]	[mm]	[mm]
Punkt I (vertikal)	2.4	4.9	15.8	2.0	4.7	20.3
Punkt I (horizontal)	-6.8	-16.4	-31.7	-3.7	-13.9	-36.3
Punkt II (vertikal)	1.1	-1.3	8.7	1.8	-0.5	12.4
Punkt II (horizontal)	-5.9	-13.0	-23.2	-3.2	-10.8	-25.1

Table E.4: Vergleich der Ergebnisse an unterschiedlichen Punkten (siehe Abbildung E.10(b)) zum Zeitpunkt $t_{fire} = 60$ min für den Gewölbequerschnitt

E.4 Fazit

Die präsentierten Simulationsergebnisse sollten einen Überblick über das Strukturverhalten von unterschiedlichen Tunnelquerschnitten (Rechteck bzw. Gewölbe) liefern. Hauptaugenmerk dieses Beitrags wurde auf den Einfluss des

Rechenmodells bzw. des Materialverhaltens gelegt. Hierbei wurden unterschiedliche Modelle zur Beschreibung des Materialverhaltens (linear-elastisch, elasto-plastisch) sowie zur Berücksichtigung der Temperaturbelastung (äquivalente Temperatur, nichtlineare Temperaturverteilung) untersucht. Des Weiteren wurde untersucht, welche Rolle die Materialparameter (f_c(T), E(T)) bzw. die gewählte Designkurve zur Berücksichtigung der temperaturabhängigen Abnahme der Materialparameter (EC2-1-2 [20] und CEB [11]) spielen.

Auf Basis der erhaltenen Ergebnisse können die folgenden Schlussfolgerungen gezogen werden:

- In einer linearen Simulation (Berücksichtigung von linear-elastischem Materialverhalten und der sogenannten äquivalenten Temperatur) werden die Zwangsschnittgrößen (insbesondere deren Spitzen) überschätzt und die Verformungen stark unterschätzt.

- Eine Mischung von elasto-plastischem Materialverhalten mit der äquivalenten Temperatur liefert zu geringe Zwangsmomente und ergibt somit eine Unterschätzung der Biegemomente. Die Verformungen sind bis zu 3-mal so groß wie bei einer linearen Simulation.

- Die nichtlineare Simulation (Berücksichtigung von elasto-plastischem Materialverhalten und der nichtlinearen Temperaturverteilung) liefert realistische Schnittgrößen für die Brandbemessung und Verformungen, die mit experimentellen Ergebnissen am besten übereinstimmen (bis zu 10-mal so groß wie bei einer linearen Simulation). Die realistische Bestimmung der Schnittgrößen ermöglicht in der Ingenieurpraxis eine wirtschaftlich optimierte Bemessung von Tragstrukturen im Brandfall.

- Eine Verwendung der Materialparameter nach CEB [11] erzeugt erheblich größeren thermischen Zwang als im Falle der Verwendung der Spannungs-Dehnungsbeziehungen aus EC 2-1-2 [20]. Daraus ergeben sich bereichsweise weitaus größere Biegemomente für die Bemessung. Laut EC 2-1-2 [20] berücksichtigen die dort enthaltenen, geringeren E-Moduli im Gegensatz zu den E-Moduli nach CEB die sogenannten "load induced thermal strains" (LITS). LITS stellt einen zusätzlichen Anteil in den Verzerrungen dar, hervorgerufen durch die kombinierte, mechanische und thermische Beanspruchung von Beton. Im EC 2-1-2 wurde dieser Einfluss indirekt auf Basis einer Vielzahl von Experimenten implementiert. Weiterführende Informationen zum Einfluss von LITS auf die Reduktion des thermischen Zwangs in Beton sind in [67] enthalten.

- Die Querschnittsform hat einen wesentlichen Einfluss auf das Strukturverhalten im Brandfall. Im Gewölbequerschnitt kommt es zufolge Temperaturbelastung zu einem Anstieg der Biegemomente im Gewölbe, welche sich gleichmäßig über den gesammten Querschnitt verteilen. Beim Rechteckquerschnitt bilden sich in den Rahmenecken Extremwerte in den Biegemomenten aus. Des Weiteren kommt es zufolge thermischen Zwangs zu einem starken Anstieg der negativen Momente bzw. zu einem Umschlagen des Moments in Deckenmitte.

E.5 Ausblick

Die Ergebnisse und Erkenntnisse aus den gezeigten und aus weiteren Struktursimulationen sollen in weiterer Folge Eingang in die Ingenieurpraxis finden. Im Rahmen der Erstellung der ÖBV-Richtlinie "Erhöhter Brandschutz mit Beton für unterirdische Verkehrsbauwerke" werden Empfehlungen zur Wahl des

Rechen- und Materialmodells in Abhängigkeit der Querschnittsform sowie des zu erreichenden Schutzniveaus bzw. des Nachweiszeitpunktes erarbeitet. So soll die Verwendung realitätsnaher Berechnungsmethoden für den Brandfall in der Ingenieurpraxis etabliert werden.

Concluding remarks & Outlook

In this work, the behavior of heated concrete as well as its effect on the load-carrying performance of underground support structures were investigated. These subjects required a wide range of research work, starting from experimental characterization of the behavior of concrete and its constituents, continuing with modeling of the encountered behavior and respective validation of the proposed models, finally yielding to simulation of the structural response by means of numerical analysis tools:

1. Material characterization

While a large amount of experimental data, which was reported in the open literature, focuses on the evolution of thermal strain of concrete under fire loading [8, 16, 35, 37, 73], experimental input for multiscale models – such as the model developed within this thesis – was still missing. Therefore, tests were conducted on cement-paste and concrete samples subjected to combined thermal and mechanical loading. In this context, a new test setup was developed, allowing simultaneous thermal and mechanical loading and, thus, providing access to the so-called load-induced thermal strain (LITS) in both the (mechanically loaded) axial direction as well as the (unloaded) radial direction. Moreover, the elastic properties (Young's modulus, Poisson's ratio) were determined as a function of mechanical and thermal load history. Finally, at the structural scale, large-scale fire experiments on concrete frames were conducted. The so-obtained data provide a profound basis for proper validation of numerical analysis tools for concrete structures subjected to fire loading.

2. Modeling

Taking into account the composite nature of concrete as well as the different behavior of its constituents (aggregates embedded in a cement-paste matrix) when subjected to temperature loading, a multiscale model was developed in this thesis. Hereby, the input data were determined from respective experiments and partially taken from the literature. The model response was validated by means of tests on heated concrete samples, focusing on the elastic and LITS properties. Finally, a novel differential LITS formulation was proposed in this thesis, capturing well the changes in mechanical loading in the course of a temperature increase – as it is the case in real-life fire scenarios.

3. Simulation

The obtained findings as regards experimental characterization and modeling were considered within a nonlinear numerical analysis tool based on the finite-element method. The underlying analysis tool allows consideration of spalling as well as the nonlinear temperature distribution within the concrete member and, hence, of the temperature-dependent mechanical response of concrete. In addition to the re-analysis of the aforementioned large-scale experiments, the performance of different tunnel cross-sections when subjected to fire loading was investigated. The developed analysis method, which was also implemented in two commercial finite-element programs used in engineering practice, was found to significantly improve the quality of results for concrete structures subjected to fire loading, especially when compared to currently-employed linear-elastic models using the so-called equivalent-temperature concept.

While the present thesis focused on the heating process of concrete and concrete structures and its effect on the load-carrying capacity of support structures, recent results of concrete subjected to combined thermal and mechanical loading – including cooling – revealed almost no effect of the applied mechanical load on the deformation behavior during cooling (see Figure 7). It is known from literature, that LITS takes place only during first heating of concrete [37], while under cooling no influence of LITS is observed. Only in case of the mechanically-unloaded concrete sample ($s = 0$ %), micro-cracks can develop

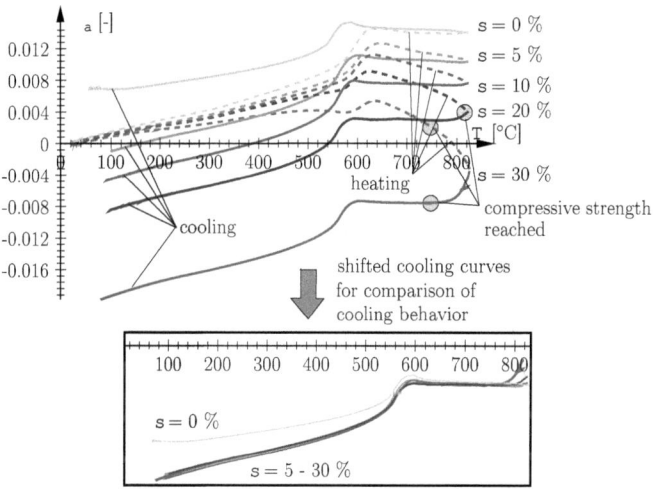

Figure 7: Heating and cooling of concrete: Evolution of axial strain as a function of temperature under constant mechanical level ($s = 100 \cdot \sigma_a/f_{c,0} = 0$ to 30 %, with the initial compressive strength $f_{c,0} = 42.4$ MPa)

due to strain incompatibilities of the concrete constituents (cement-paste and aggregates), leading to different strain behavior during heating as well as cooling. Accordingly, the strain behavior of concrete during cooling seems to be almost independent of the level of mechanical loading, except for the case of unloaded concrete. In addition to mechanical loading, temperature gradients within the cylindrical concrete samples (between surface and core) lead to additional (eigen-) stresses, with the total stress reaching the compressive strength of concrete (see response of concrete samples for $s = 30$ % in Figure 8). This resulted in additional deformations, taking place at the beginning of the cooling process.

Whereas the developed model captures well the material response (both heating and cooling) for stress situations found below the compressive strength, the additional (plastic) deformation caused by loading exceeding the compressive strength is not yet taken into account, explaining the discrepancy observed in Figure 8.

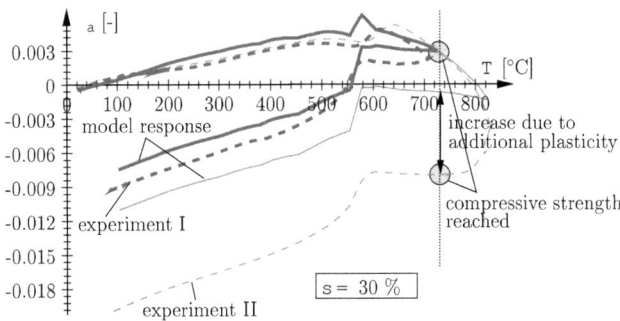

Figure 8: Evolution of axial strain: Experimental observation vs. model response for two experiments ($s = 30$ %)

Accordingly, modeling the behavior of concrete at high temperatures – as presented in this thesis – can be further improved by focusing on

- the concrete behavior when loaded beyond the compressive strength, becoming relevant during cooling of concrete when LITS is disregarded,
- the consideration of the microcrack density within the proposed multiscale model for heated concrete, finally explaining the changes in deformation behavior as well as the temperature-dependent mechanical properties by the

146

development and suppression (by mechanical load) of microcracks within the cement-paste matrix.

As regards the presented numerical analysis tool for the simulation of the response of underground support structures subjected to fire loading, the results presented in this thesis are currently providing the basis for a new Austrian guideline to be published by the Austrian Society for Construction Technology, entitled "Increased fire protection with concrete for underground structures". This guideline will contain recommendations regarding the proper choice of the underlying numerical and material model to be used for the simulation of underground structures subjected to fire loading. In this context, the outcome of this thesis contributes to the improvement of the quality of results by the use of more realistic analysis tools which, finally, will lead to a change of engineering practice regarding the design of concrete structures subjected to fire.

Bibliography

[1] M.S. Abrams. Compressive strength of concrete at temperatures to 1600 F. *ACI-SP 25*, pages 33–58, 1971.

[2] E. Aigner, R. Lackner, and Ch. Pichler. Multiscale prediction of viscoelastic properties of asphalt concrete. *Materials in Civil Engineering (ASCE)*, 21(12):771–780, 2009.

[3] L. Alarcon-Ruiz, G. Platret, E. Massieu, and A. Ehrlacher. The use of thermal analysis in assessing the effect of temperature on a cement paste. *Cement and Concrete Research*, 35:609–613, 2005.

[4] A. Amouzandeh. *Development and application of a computational fluid dynamics code to predict the thermal impact on underground structures in case of fire*. PhD thesis, Vienna University of Technology, Vienna, Austria, 2012.

[5] A. Amouzandeh, M. Zeiml, and R. Lackner. Design of a CFD-code to predict the thermal impact of fires in underground structures - Part I: analysis of sub models. *Fire Safety Journal*, 2012. Submitted for publication.

[6] A. Amouzandeh, M. Zeiml, and R.. Lackner. Design of a CFD-code to predict the thermal impact of fires in underground structures - Part II: re-analysis of full-scale fire tests. *Fire Safety Journal*, 2012. Submitted for publication.

[7] Y. Anderberg. Spalling phenomena in hpc and oc. In L.T. Phan, N.J. Carino, D. Duthinh, and E. Garboczi, editors, *Proceedings of the International Workshop on Fire Performance of High-Strength Concrete*, pages 69–73, Gaithersburg, Maryland, 1997. NIST.

[8] Y. Anderberg and S. Thelandersson. Stress and deformation characteristics of concrete at high temperatures: 2. experimental investigation and material behaviour model. Technical Report 54, Lund Institute of Technology, Lund, 1976.

[9] S. Bauer and R. Lackner. Discontinuity layout optimization in upscaling of effective strength properties in matrix-inclusion materials. In *Proceedings of Complas XI*, Barcelona, Spain, 2011.

[10] Brite Euram III BRPR-CT95-0065. Understanding and industrial application of High Performance Concrete in High Temperature Environment - Final Report. Technical report, HITECO, 1999.

[11] CEB. *Fire Design of Concrete Structures, Bulletin d'Information 208*. CEB, Lausanne, Switzerland, 1991.

[12] G.R. Consolazio, M.C. McVay, and J.W. Rish III. Measurement and prediction of pore pressure in saturated cement mortar subjected to radiant heating. 95(5):525–536, 1998.

[13] C.R. Cruz and M. Gillen. Thermal Expansion of Portland Cement Paste, Mortar, and Concrete at High Temperatures. *Fire and Materials*, 4(2), 1980.

[14] M.J. DeJong and F.-J. Ulm. The nanogranular behavior of C-S-H at

elevated temperatures (up to 700 °C). *Cement and Concrete Research*, 37:1–12, 2007.

[15] J.Dweck, P.F.Ferrerira da Silva, P.M.Büchler, and F.K. Cartledge. Study by thermogravimetry on the evolution of ettringite phase during type II portland cement hydration. *Journal of Thermal Analysis and Calorimetry*, 69:179–186, 2002.

[16] C. Ehm. *Versuche zur Festigkeit und Verformung von Beton unter zweiaxialer Beanspruchung und hohen Tempearturen [Experiments on strength and strain of concrete under biaxial loading at high temperatures]*. PhD thesis, University of Braunschweig, Braunschweig, Germany, 1985.

[17] I.A.El-Arabi, H.Duddeck, and H.Ahrens. Structural analysis for tunnels exposed to fire temperatures. *Tunneling and Underground Space Technology*, 7(1):19–24, 1992.

[18] EN 1991-1-2. *Eurocode 1 – Einwirkungen auf Tragwerke – Teil 1-2: Allgemeine Regeln – Brandeinwirkungen auf Tragwerke [Actions on structures – Part 1-2: General actions – Actions on structures exposed to fire]*. European Committee for Standardization (CEN), 2011.

[19] EN 1992-1-1. *Eurocode 2 – Bemessung und Konstruktion von Stahlbeton- und Spannbetontragwerken – Teil 1-1: Allgemeine Bemessungsregeln und Regeln für den Hochbau [Eurocode 2 – Design of concrete structures – Part 1-1: General rules and rules for buildings]*. European Committee for Standardization (CEN), 2009.

[20] EN 1992-1-2. *Eurocode 2 – Bemessung und Konstruktion von Stahlbeton- und Spannbetontragwerken – Teil 1-2: Allgemeine Regeln – Tragwerksbemessung für den Brandfall [Eurocode 2 – Design of concrete structures –*

Part 1-2: General rules – Structural fire design]. European Committee for Standardization (CEN), 2007.

[21] C. Feist, M. Aschaber, and G. Hofstetter. Numerical simulation of the structural behavior of RC tunnel structures subjected to fire. In *Proceedings of ECCOMAS Thematic Conference on Computational Methods in Tunneling (EURO:TUN 2007)*, Vienna, Austria, 2007.

[22] J. Füssl and R. Lackner. Multiscale fatigue model for bituminous mixtures. *International Journal of Fatigue*, 33:1435–1450, 2011.

[23] J. Füssl, R. Lackner, J. Eberhardsteiner, and H. A. Mang. Failure modes and effective strength of two-phase materials determined by means of numerical limit analysis. *Acta Mechanica*, 195:185–202, 2008.

[24] A. Galek, H. Moser, T. Ring, M. Zeiml, J. Eberhardsteiner, and R. Lackner. Mechanical and transport properties of concrete at high temperatures. In *Applied Mechanics and Materials*, volume 24-25, pages 1–11. Trans Tech Publications, Switzerland, 2010.

[25] D. Gawin, F. Pesavento, and B. A. Schrefler. Modelling of deformations of high strength concrete at elevated temperatures. 37:218–236, 2004.

[26] H. Hejny. Work Package 4 – Fire effects and tunnel performance: system structural response. Technical Report 736, UPTUN – UPgrading of existing TUNnels, 2008.

[27] H. Ingason, S. Gustavsson, and M. Dahlberg. Heat Release Rate Measurements in Tunnel Fires. Technical Report 08, SP Swedish National Testing and Research Institute, Sweden, 1994.

[28] H. Ingason and A. Lönnermark. Heat release rates from heavy goods vehicle trailer fires in tunnels. *Fire Safety Journal*, 40:646–668, 2005.

[29] A. Jäger and R. Lackner. Identification of viscoelastic model parameters by means of cyclic nanoindentation testing. *International Journal of Material Research*, 99:829–835, 2008.

[30] A. Jäger and R. Lackner. Finer-scale extraction of viscoelastic properties from nanoindentation characterised by viscoelasticplastic response. *Strain*, 45:45–54, 2009.

[31] A. Jäger, R. Lackner, and J. Eberhardsteiner. Identification of viscoelastic properties by means of nanoindentation taking the real tip geometry into account. *Meccanica*, 42 (3):293–306, 2007.

[32] R. Jansson. *Material properties related to fire spalling of concrete*. PhD thesis, Lund Institute of Technology, Lund, Sweden, 2008.

[33] A.H. Jay. The thermal expansion of quartz by X-ray measurements. *Proceedings of the Royal Society of London*, pages 227–247, 1933.

[34] G. Khoury and C.E. Majorana. Thermo-hydral behaviour. In G. Khoury and C.E. Majorana, editors, *Effect of Heat on Concrete*, pages 1–18, Udine, 2003. International Centre for Mechanical Science.

[35] G.A. Khoury. Strain components of nuclear-reactor-type concretes during first heat cycle. *Nuclear Engineering and Design*, 156:313–321, 1995.

[36] G.A. Khoury, B.N. Grainger, and P.J.E. Sullivan. Strain of concrete during first heating to 600°C. *Magazine of Concrete Research*, 37 (133):195–215, 1985.

[37] G. A. Khoury, B. N. Grainger, and P. J. E. Sullivan. Transient thermal strain of concrete: literature review, conditions within specimen and behaviour of individual constituents. *Magazine of Concrete Research*, 37(132):131–144, 1985.

[38] T. Kührer. Nachbrandfestigkeit von zementgebundenen Werkstoffen. Druckversuche und Thermogravimetriemessungen. Technical report, TU Wien, 2008.

[39] W. Kusterle, W. Lindlbauer, G. Hampejs, A. Heel, P.-F. Donauer, M. Zeiml, W. Brunnsteiner, R. Dietze, W. Hermann, H. Viechtbauer, M. Schreiner, R. Vierthaler, H. Stadlober, H. Winter, J. Lemmerer, and E. Kammeringer. Brandbeständigkeit von Faser-, Stahl- und Spannbeton [Fire resistance of fiber-reinforced, reinforced, and prestressed concrete]. Technical Report 544, Bundesministerium für Verkehr, Innovation und Technologie, Vienna, 2004. In German.

[40] R. Lackner and H. A. Mang. Scale transition of steel-concrete interaction I: Model. *ASCE*, 129(4):389–402, 2003.

[41] R. Lackner and H. A. Mang. Scale transition of steel-concrete interaction II: Applications. *ASCE*, 129(4):403–413, 2003.

[42] R. Lackner, Ch. Pichler, and A. Kloiber. Artificial ground freezing of fully saturated soil: Viscoelastic behavior. *Journal of Engineering Mechanics (ASCE)*, 134(1):1–11, 2008.

[43] D. L. Lakshtanov, S. V. Sinogeikin, and J. D. Bass. High-temperature phase transitions and elasticity of silica polymorphs. *Physics and Chemistry of Minerals*, 34(1):11–22, 2007.

[44] J. Lee, Y. Xi, K. Willam, and Y. Jung. A multiscale model for modulus of elasticity of concrete at high temperatures. *Cement and Concrete Research*, 39:754–762, 2009.

[45] T. Lemaire and Y. Kenyon. Large Scale Fire Tests in the second Benelux Tunnel. *Fire Technology - Springer Science*, 42:329–350, 2006.

[46] A. Lönnermark and H. Ingason. Gas temperatures in heavy goods vehicle fires in tunnels. *Fire Safety Journal*, 40:506–527, 2005.

[47] H. Lun, R. Lackner, A. Galek, H. Moser, and M. Zeiml. Ergebnisse der Residual- und Heißpermeabilitätsversuche und Vergleich mit den Ergebnissen der MIP. Technical report, Innsbruck, Austria, 2012. In German.

[48] MARC Analysis Research Corporation. *Volume C - Program Input*, 1994. Manual.

[49] T. Mori and K. Tanaka. Average stress in matrix and average elastic energy of materials with misfitting inclusions. *Acta Metallurgica*, 21:571–574, 1973.

[50] C. V. Nielsen, C. J. Pearce, and N. Bićanić. Improved phenomenological modelling of transient thermal strains for concrete at high temperatures. *Computers and Concrete*, 1(2):189–209, 2004.

[51] P. Nischer, J. Steigenberger, and H. Wiklicky. Praxisverhalten von erhöht brandbeständigem (Innenschalen-) Beton (EBB) [Practical use of fire resistant concrete for tunnel linings]. Technical Report, FFF-project no. 806201, Forschungsinstitut der Vereinigung der österreichischen Zementindustrie (VÖZ), Vienna, 2004. In German.

[52] ÖBV-Richtlinie Wasserundurchlässige Betonbauwerke – Weiße Wannen.

Österreichische Vereinigung für Beton- und Bautechnik (ÖVBB), 2009. In German.

[53] Österreichisches Normungsinstitut. *Beton – Teil 1: Festlegung, Herstellung, Verwendung und Konformitätsnachweis [Concrete – Part 1: Specification, production, use and verification of conformity]*, 2004. In German.

[54] M. Petkovski. Effects of stress during heating on strength and stiffness of concrete at elevated temperature. *Cement and Concrete Research*, 40:1744–1755, 2010.

[55] B. Pichler and C. Hellmich. Upscaling quasi-brittle strength of cement paste and mortar: A multi-scale engineering mechanics model. *Cement and Concrete Research*, 41:467–476, 2011.

[56] Ch. Pichler. *Multiscale characterization and modeling of creep and autogenous shrinkage of early-age cement-based materials*. PhD thesis, Vienna University of Technology, Juni 2007.

[57] Ch. Pichler and R. Lackner. A multiscale creep model as basis for simulation of early-age concrete behavior. *Computers and Concrete*, 5(4):295–328, 2008.

[58] Ch. Pichler and R. Lackner. A multiscale micromechanics model for early-age basic creep of cement-based materials. *International Journal of Computers and Concrete*, 5(4):295–328, 2008.

[59] Ch. Pichler and R. Lackner. Identification of logarithmic-type creep of calcium-silicate-hydrates by means of nanoindentation. *Strain*, 45(1):17–25, 2009.

[60] Ch. Pichler and R. Lackner. Upscaling of viscoelastic properties of highly-filled composites. Investigation of matrix-inclusion-type morphologies with power-law viscoelastic material response. *Composites Science and Technology*, 69:2410–2420, 2009.

[61] Ch. Pichler, R. Lackner, and E. Aigner. Generalized self-consistent scheme for upscaling of viscoelastic properties of highly-filled matrix-inclusion composites – Application in the context of multiscale modeling of bituminous mixtures. *Composites Part B: Engineering*, 43(3):457–464, 2012.

[62] Ch. Pichler, R. Lackner, and H.A. Mang. Multiscale model for creep of shotcrete – From logarithmic-type viscous behavior of CSH at the μm-scale to macroscopic tunnel analysis. *Journal of Advanced Concrete Technology*, 6(1):91–110, 2008.

[63] Ch. Pichler, G. Metzler, Ch. Niederegger, and R. Lackner. Thermomechanical optimization of porous building materials based on micromechanical concepts: Application to load-carrying insulation materials. *Composites Part B: Engineering*, 43(3):1015–1023, 2012.

[64] E. Richter and D. Hosser. Baulicher Brandschutz bei Verkehrstunneln in offener Bauweise [Fire protection for tunnels]. *Beton- und Stahlbetonbau*, 97(4):178–184, 2002. In German.

[65] T. Ring, C. Wikete, H. Kari, M. Zeiml, and R. Lackner. Der Einfluss des Rechen- und Materialmodells auf die Strukturantwort bei der Simulation von Tunnel unter Brandbelastung. *Bauingenieur*, 2012. Under preparation.

[66] T. Ring, M. Zeiml, and R. Lackner. Abschlussbericht Brandversuche zum Abplatz- und Strukturverhalten von Tunnel mit Rechtecksquerschnitt.

Technical report, Vienna University of Technology – KIRAS Project, Vienna, Austria, 2012. In German.

[67] T. Ring, M. Zeiml, and R. Lackner. Thermo-mechanical behavior of concrete at high temperature: From micromechanical modeling towards tunnel safety assessment. *Cement and Concrete Research*, 2012. Submitted to Journal.

[68] T. Ring, M. Zeiml, and R. Lackner. Underground concrete frame structures subjected to fire loading: Part I – large scale fire tests. *Engineering Structures*, 2012. Accepted under revision.

[69] T. Ring, M. Zeiml, and R. Lackner. Underground concrete frame structures subjected to fire loading: Part II – re-analysis of large scale fire tests. *Engineering Structures*, 2012. Accepted under revision.

[70] T. Ring, M. Zeiml, R. Lackner, and J. Eberhardsteiner. Experimental investiagtion of strain behavior of heated cement paste and concrete. *Strain*, 2012. Accepted under revision.

[71] K. Savov, R. Lackner, and H.A. Mang. Stability assessment of shallow tunnels subjected to fire load. *Fire Safety Journal*, 40:745–763, 2005.

[72] U. Schneider. *Zur Kinetik festigkeitsmindernder Reaktionen in Beton bei hohen Temperaturen*. PhD thesis, TU Braunschweig, Braunschweig, Germany, 1973. In German.

[73] U. Schneider. Concrete at high temperature – a general review. *Fire Safety Journal*, 13:55–68, 1988.

[74] M.J. Terro. Numerical modeling of the behavior of concrete structures in fire. *ACI Structural Journal*, 95(2):183–193, 1998.

[75] S. Thelandersson. Modeling of combined thermal and mechanical action in concrete. *Journal of Engineering Mechanics (ASCE)*, 113(6):893–906, 1987.

[76] K.-C. Thienel. Festigkeit und Verformung von Beton bei hoher Temperatur und biaxialer Beanspruchung – Versuche und Modellbildung [Strength and deformation of concrete at high temperature – experiments and modeling]. Technical Report 437, Deutscher Ausschuss für Stahlbeton, Berlin, 1994.

[77] S. Tsivilis, G. Kakali, E. Chaniotakis, and A. Souvaridou. A study on the hydration of portland limestone cement by means of TG. *Journal of Thermal Analysis*, 52:863–870, 1998.

[78] J.W ageneder. Traglastuntersuchungen unter Brandeinwirkungen [Ultimate load investigations considering fire load]. *Bauingenieur*, 77:184–192, April 2002. In German.

[79] M. Zeiml. *Concrete subjected to fire loading - From experimental investigation of spalling and mass-transport properties to structural safety assessment of tunnel linings under fire*. PhD thesis, Vienna University of Technology, Juli 2008.

[80] M. Zeiml, R. Lackner, D. Leithner, and J. Eberhardsteiner. A novel experimental technique for determination of the permeability of concrete subjected to high temperature. *Cement and Concrete Research*, 38:699–716, 2008.

[81] M. Zeiml, R. Lackner, F. Pesavento, and B.A. Schrefler. Thermo-hydro-chemical couplings considered in safety assessment of shallow tunnels subjected to fire load. *Fire Safety Journal*, 43(2):83–95, 2008.

[82] M. Zeiml, D. Leithner, R. Lackner, and H.A. Mang. How do polypropylene fibers improve the spalling behavior of in-situ concrete? *Cement and Concrete Research*, 36(5):929–942, 2006.

[83] M. Zeiml, T. Ring, and R. Lackner. Influence of spalling and thermomechanical loading on the structural performance of tunnel linings under fire. In *Proceedings of the 1st International Conference on Computational Technologies in Concrete Structures (CTCS'09)*, page 256, Korea, 2009. Techno Press.

[84] Y. Zhang, R. Lackner, M. Zeiml, and H.A. Mang. Spalling risk assessment of concrete subjected to different fire scenarios. In *Proceedings of the 2nd International Conference on Microdurability*, Amsterdam, 2012.

[85] Y. Zhang, C. Pichler, Y. Yuan, M. Zeiml, and R. Lackner. Micromechanics-based multifield framework for early-age concrete. *Engineering Structures*, 2012. Accepted for publication.

i want morebooks!

Buy your books fast and straightforward online - at one of world's fastest growing online book stores! Environmentally sound due to Print-on-Demand technologies.

Buy your books online at
www.get-morebooks.com

Kaufen Sie Ihre Bücher schnell und unkompliziert online – auf einer der am schnellsten wachsenden Buchhandelsplattformen weltweit! Dank Print-On-Demand umwelt- und ressourcenschonend produziert.

Bücher schneller online kaufen
www.morebooks.de

VDM Verlagsservicegesellschaft mbH
Heinrich-Böcking-Str. 6-8 Telefon: +49 681 3720 174 info@vdm-vsg.de
D - 66121 Saarbrücken Telefax: +49 681 3720 1749 www.vdm-vsg.de

Printed by Books on Demand GmbH, Norderstedt / Germany